PRODUCT SENSE

ENGINEERING YOUR *IDEAS* INTO *REALITY*

A MUST READ FOR ENTREPRENEURS & INVENTORS

DR. K.R. SURESH NAIR

ISBN 979-8-89277-847-3

Table of Contents

Acknowledgements

The society as a whole has contributed towards my learning and I have been thinking of ways to give it back, and came the idea of writing a book. Of course, I am familiar with writing technical articles, but too much of technical will lead to readers skipping and turning the pages fast. Finally decided to come up with a book which I believe is readable and enjoyable by anyone. I have been meeting young entrepreneurs almost every day, enjoy listening to their innovations and possibly mentor them on need basis. My regular visit to many of the Incubation centers at IITs and like prestigious institutions and subsequent discussions with startups compelled me to pen down the knowledge and experience I gained.

All the achievements mentioned in this book are due to the hard work of my team members and my sincere appreciation and acknowledgements to my colleagues and senior management at TIFR-SAMEER-R&D laboratory of Government of India, NeST Group, Social Alpha, Amara Raja Design Alpha and Biophoton Technologies.

I am habitual of taking professional risks in my career and this was possible only with the rock support of my spouse Dr. Geetha Nair, an ENT Surgeon. She has been a positive critic herself which made me move cautiously. My focused discussions with my son Akhil Suresh, a successful entrepreneur himself made me tuned to new developments and adapt towards new business environments. The love and support by Akhil, Sana and Kiara make me grow younger as years progress.

My heartfelt remembrances and prayers to my late parents for their love and patiently waiting for many years to complete my education. I thank my sisters on feeling always proud on me. And late parents-in-law who remained strong force to my professional growth.

Special mention to Sri. S.Vijaya Kumar, who placed the seed and nurtured the ambition of becoming a Scientist/researcher in me and motivated every step forward.

Thanks to Notion Publishers to accept and publish this book.

Foreword

These days, 'innovation' is a word that gets tossed around a lot. It seems like everyone wants to be the next big inventor with an idea that will change the world. But there's more to it than just having a good idea. What's often missed is the need for patience and discipline to turn an idea into a commercially viable and scalable product.

Dr. Suresh Nair stands out as a paragon in this relentless pursuit of innovation. With a distinguished academic background from a premier global institution, an illustrious career in multinational corporations, and a track record of spearheading R&D teams, Dr. Nair's contributions are substantial. His portfolio boasts numerous patents, and he has successfully elevated a company to new heights through strategic partnerships with established firms. I felt compelled to pen this foreword, aware that Dr. Nair's humility might prevent him from sharing these accolades. However, it is essential for you, the reader, to comprehend the breadth of his experience, which imbues this book with unparalleled insight and invaluable wisdom.

The truest legacy one might aspire to leave is the conversion of vision into reality. To embark on this journey is to embrace the full scope of the endeavor, armed with discipline, an eagerness for rapid learning, the courage to step beyond one's comfort zone, and the resilience to persevere through economic and societal challenges.

I'm not only his son but his greatest admirer. I've seen him help shape people and companies over the years, and I've seen the tough problems he had to solve. This includes creating important devices that have saved lives. This book shares his vision and experiences so that more people can learn from him and carry on his work of turning great ideas into products that make a difference.

– Akhil Suresh Nair

CEO, Xena Intelligence

Prologue

In the early dawn of my career, nestled within the hallowed halls of Government Research and Development, I embarked on a journey that was nothing short of a dream for any Scientist or Engineer. My days were consumed with the intricate design of an atmospheric radar – a task both prestigious and exhilarating. Surrounded by an excellent working ambiance, where concessional healthy food fueled our bodies and a conducive environment nourished our minds, I delved deep into the realms of analog, digital, RF and Microwave designs, software algorithms, and simulation techniques.

Being an integral part of the system design team, I absorbed knowledge like a sponge, learning nuances and complexities that only such hands-on experience could offer. The radar system, our brainchild, was eventually installed. It performed splendidly, surpassing all expectations. Many years later, a nostalgic journey took me back to that remote area, and to my amazement, I found that the facility had grown exponentially, now boasting an array of atmospheric experimental facilities – a testament to our foundational work.

After this fulfilling endeavor, a new challenge beckoned – research on advanced optical chip designs. Supported generously by government funding, we established a design and engineering center dedicated to integrated optics. As the leader of this venture, I recall this term with a sense of pride and fondness. It was a period marked not only by challenges but also by the relentless dedication of a team of young Engineers. Together, we toiled day and night, our efforts culminating in the successful demonstration of lithium niobate modulators and Glass based 1x8 power splitters in optical communication wavelengths.

This triumph caught the attention of an optical company from the US, who were so impressed by our innovations that they decided to acquire the technology we developed. That period in my career remains etched in my memory, a time when hard work, team spirit, and groundbreaking achievements intertwined to create something truly remarkable.

As I journeyed through my career, a pivotal moment arrived, steering me from the prestigious corridors of R&D into the vibrant, unpredictable world of private industry. This transition was sparked by a US company, whose manufacturing operations were nestled in the lush landscapes of Kerala. Their endorsement was the catalyst for my shift, a leap from the familiar realms of a celebrated research organization into the exhilarating uncertainty of a private enterprise.

The move was met with whispers of caution and trepidation. The industry, many said, was a different beast – unpredictable, with the ever-looming shadow of job insecurity. This decision wasn't mine alone; it bore a significant impact on my family. My wife, an accomplished ENT surgeon, had to close the doors to her own clinic, stepping into an uncertain future without a confirmed job in sight. My son, in the midst of his academic journey, faced the daunting task of leaving his classmates behind to join a new school, one that graciously allowed a mid-term admission.

This chapter of my life was more than a career shift; it was a deep dive into the world of global business. It was here that I learned the art of being an opening batsman in client meetings, skillfully pitching for business right from the first encounter. I was tasked with leading a larger R&D team, a group of brilliant minds racing against time to deliver products that were not only innovative but also ready for the rigors of manufacturing.

Amidst these challenges, an exhilarating opportunity surfaced. A global network provider laid down the gauntlet – an open challenge to conceptualize Fiber to Home based systems. The stakes were high; a multimillion-dollar business awaited us if we could deliver a concept that was innovative, impressive, better in specifications, and cost-effective.

Preparation for this challenge was intense. As the opening batsman, I knew the pitch demanded excellence that had not been achieved by anyone before. My team and I embarked on an exhaustive journey of literature surveys and meticulous studies. We delved into every nuance, every possibility, exploring uncharted territories to bring this groundbreaking concept to life.

This period was more than just a phase of professional growth; it was a transformative experience that reshaped my understanding of business, innovation, and leadership. It was a testament to the power of resilience, adaptability, and the relentless pursuit of excellence, even when faced with the unknown. This journey, rife with challenges and triumphs, not only charted a new course for my career but also left an indelible mark on my personal growth, teaching me invaluable lessons about the complexities and rewards of the global business landscape.

At this crucial juncture of my career, I found myself at a crossroads, grappling with two daunting options, each fraught with its own set of risks and potential rewards. The task ahead was Herculean – to meet a specification in fiber optics that had eluded the best in the field, a feat not yet documented in the annals of technological advancements. The decision I was about to make would not only define the trajectory of our project but could also seal the fate of potential business opportunities.

The first option was to play it safe – to inform the committee that we could achieve what had been documented in literature to date, but not the advanced specifications they demanded. This approach, however pragmatic, ran the risk of the committee perceiving it as an admission of our limitations, a confession that we were not equipped to meet their futuristic standards.

The alternative was more audacious – to commit to the demanding specifications, despite the knowledge that no entity had yet achieved them. This path was equally perilous, as it could invite skepticism regarding our understanding and experience in the technology, casting doubts on our capability to deliver.

As the meeting loomed closer, sleep became elusive. The weight of the decision hung heavily on my mind – it was a crime to lose such a monumental business opportunity, especially when we were relatively new to this realm of technology. It was during one of these restless nights that a spark of an idea ignited in my mind, a strategy that could potentially turn the tide in our favor.

In the meeting, I took a leap of faith. I declared our commitment to achieve the demanding specifications, despite the absence of any precedent. I posited a belief – if such an established global company had set these specifications, they must be within the realm of possibility. It was a bold stance, but one that was rooted in a deep-seated belief in the art of the possible.

This audacious commitment paid off. We were given the opportunity to demonstrate our product within a few months. My team, fueled by determination and ingenuity, embarked on a series of improvisations. After tireless efforts and numerous iterations, we succeeded in developing a concept that met the demanded specifications – a feat that had seemed nearly impossible.

Thus, we not only secured the global network provider as our client but also ventured into the next phase of challenges – realizing this breakthrough in volume manufacturing with specifications that were hitherto unrealized. The ensuing months were a whirlwind of research, development, and relentless work, culminating in the establishment of a manufacturing facility that turned our groundbreaking concept into a tangible reality.

Post the exhilarating journey of pioneering fiber optics technology, life settled into a rhythm filled with presentations, product deliveries, and the continuous pursuit of adding new clients from diverse sectors. Innovation, once a mere aspect of my job, had now transformed into an addiction, a relentless drive that reshaped my vision of the future. I realized that the true essence of enjoying this creative fervor was to step into the world of entrepreneurship.

Venturing into entrepreneurship in my fifties presented a unique scenario. The startup landscape was predominantly a playground for the young and daring minds. Yet, here I was, embracing this sweet challenge with open arms, ready to rewrite the rules. I immersed myself in the world of startups, participating in competitions, and absorbing wisdom through a series of short courses at top universities, all focused on the art of entrepreneurship. As I navigated this new path, I found a curious truth unveiling itself – with each year spent as an entrepreneur, I felt as though I was journeying back to my forties. It was as if entrepreneurship had the power to reverse the sands of time, rejuvenating the spirit with each innovative step.

The first imperative as an entrepreneur was clear – to conceive innovative ideas. During this quest, I chanced upon a picture in a book that portrayed Adam and Eve, Isaac Newton, and Steve Jobs, all linked by the motif of an apple, each symbolizing revolutionary changes in their respective eras. This imagery sparked a thought – being in a land abundant with coconuts, a Newtonian revelation might end rather abruptly for me. I chuckled at the thought and decided against the idea of replicating such epiphanies.

Instead, I turned my focus to understanding customer pain points. The goal was no longer about emulating historical figures or their eureka moments. It was about carving a unique niche, identifying areas where innovation could alleviate challenges, and enhance the customer experience. This shift in perspective marked the beginning of a journey that was as much about unlearning as it was about learning – a journey where age was merely a number, and the true metric of vitality was the ability to innovate and transform the ordinary into the extraordinary. As an entrepreneur, every day was a new canvas, waiting to be painted with ideas that could potentially change the world, one innovation at a time.

Now, let's transition from personal ambitions and narrative to delve into a larger movement. We find ourselves at the cusp of a new era. At the dawn of the twenty-first century, a revolutionary wave began to reshape the commercial landscapes of the world. This was not a mere shift in business paradigms, but a radical reimagining of what it means to create, innovate, and disrupt. Today, as we stand amidst the echoes of this transformation, we find ourselves in the contemporary age of startup aspirations—a period marked not only by a surge in entrepreneurial spirit but by the birth of enterprises that embody the very spirit of disruption and innovation.

In this dynamic era, startups are far more than just business entities. They are the vanguard of change, challenging the status quo and offering ingenious solutions to some of society's most persistent issues. The fuel for this transformative fire has been a combination of rapidly advancing technology, the globalization of digital connectivity, and a consumer base that is more informed and engaged than ever before.

Technology, once the gatekeeper to industrial power, has now become the great leveler. The proliferation of digital technologies has dismantled traditional barriers to entry, enabling startups to emerge and scale with unprecedented speed. The digital revolution has empowered entrepreneurs everywhere with access to the tools necessary to build and deploy complex solutions, from the use of artificial intelligence in personalizing customer experiences to leveraging the Internet of Things (IoT) for smarter cities.

The startup ecosystem's growth has been bolstered significantly by venture capital, which has provided the necessary resources for expansion and scale. The surge of investment in startup ventures speaks to a belief in the power of innovation—a belief that is as financial as it is ideological. Moreover, the role of incubators and accelerators has been pivotal, creating environments ripe for innovation and providing the mentorship, resources, and networking opportunities that are vital for nascent startups.

The leap into the future is being powered by technologies like AI, blockchain, and IoT, propelling startups towards creating solutions that were once the realm of science fiction. In an age where collaboration knows no borders, the rise of remote operations has allowed startups to tap into global talent pools and markets, fostering a culture of ubiquitous connectivity.

The narrative takes a compelling turn as we pivot to India—a land where ancient traditions meet modern ambitions. India's startup ecosystem is burgeoning, marked by an eclectic mix of sectors that mirror the country's diversity, from fintech and e-commerce to health tech and ed-tech.

With over 55,000 startups launched and 40,000 active startups, India's startup landscape is vibrant and teeming with potential. Of these, 33,000 are recognized by the Department for Promotion of Industry and Internal Trade (DPIIT), showcasing the breadth of the sector. The journey from ideation to realization is marked by robust support, with 385 funded startups and a staggering $63 billion raised by them, making clear the substantial financial backing that propels these ventures. The ecosystem has been a fertile ground for innovation, with 5,400 funding deals witnessed.

One cannot overlook the demographic dividend—a vast, youthful population that not only drives innovation but also amplifies consumption. This demographic strength, coupled with the rapid adoption of digital technologies among the populace, has positioned India as a global powerhouse of startup innovation. Initiatives like 'Make in India', 'Digital India' and 'Atma Nirbhar' have further augmented this growth, propelling India towards becoming a global startup hub.

Consider the case of Paytm, which stands as a testament to the transformative potential of the Indian market. What began as a humble mobile recharge application has now metamorphosed into a comprehensive financial services colossus, demonstrating the agility and adaptability of Indian startups. Paytm's journey—from facilitating simple transactions to providing a full spectrum of financial services—mirrors the country's own trajectory of growth and innovation.

The impact of startups is profound and all-encompassing. These burgeoning enterprises have entered with not just new products or services, but with bold new visions that challenge long-standing industry giants and redefine the essence of value creation.

The Metamorphosis of Markets

For these startups, innovation serves as the guiding star, propelling them toward new frontiers brimming with potential. Their influence on today's market dynamics cannot be overstated. We're witnessing a shift in focus from traditional business offerings to the crafting of comprehensive experiences and solutions. These startups are not just filling existing gaps; they are adept at identifying latent needs, thereby creating entirely new markets that were previously unimaginable.

Industry Disruption through Innovation

Every industry touched by this wave of innovation has stories of transformation to tell:

Fintech: In the financial sector, startups like Square and Stripe have shattered the traditional gatekeeping of financial services. They've made these services widely accessible, enabling small businesses and individuals to engage in financial activities that were once the sole domain of large banks.

Healthcare: Startups such as Zocdoc and 23andMe have pivoted healthcare from a generic, provider-centric model to one that empowers patients. They've brought forward a new era of personalized medicine, where treatments and health strategies are tailored to the individual, not the masses.

Transportation: Disruptive entities like Uber and Lyft have not only redefined personal mobility but also significantly influenced urban planning and the very notion of car ownership, presenting alternatives that are flexible and on-demand.

Driving Factors Behind Disruptive Innovation

Several factors fuel the engine of disruptive innovation within startups:

Technological Leverage: Startups are adept at harnessing new technologies to disrupt existing business models or to forge new ones from the ground up.

Agility and Adaptability: These new ventures are defined by their flexibility, enabling them to navigate market feedback and technological changes swiftly and efficiently.

Customer-Centric Approach: Success for many startups hinges on their commitment to the customer experience, tailoring their offerings to the precise needs of distinct market niches.

Risk Tolerance: Startups often court risks that more established firms avoid, embracing the potential for groundbreaking innovation.

The narrative of disruption and innovation finds a particularly resonant chord in India. The Indian startup ecosystem has been a catalyst for substantial shifts across various industries. In India, startups have played a critical role in closing the divide between digital possibilities and physical realities. By innovating in agritech, they have introduced technology to agriculture, enhancing productivity and sustainability for farmers. In the realm of education, EdTech startups have delivered educational resources to remote areas, fundamentally changing the landscape of learning and opportunity.

Continuing from the stirring narrative of startups redefining the market landscape through innovative practices, we delve into the essence of what makes an innovation truly successful. The journey from a spark of creativity to a commercially viable product is both complex and enlightening. An idea, irrespective of its origin, holds potential only when it resonates with the needs and interests of the market and makes commercial sense. For entrepreneurs, this journey involves nurturing an idea—often stemming from their education, expertise, or social background—and meticulously translating it into a tangible product or service.

In the bustling ecosystem of startups, 'Innovation' has become a revered concept, especially among millennials, many of whom dream of launching ventures that disrupt the status quo. To support such aspirations, numerous organizations have emerged, offering incubation services to shepherd these ideas to fruition. Yet, despite the support, the market hasn't witnessed a proportionate surge in innovative products. This presents a critical moment to introspect the reasons behind this imbalance, to identify the gaps in the innovation process, and to pivot towards nurturing quality rather than quantity.

The challenge often lies in the evaluation of these innovative proposals. Many proposals, despite their impressive presentation, are derived from existing international research papers or are iterations of products already available elsewhere. A common thread in these proposals is the promise of reduced costs, usually based on an estimated Bill of Materials (BOM). However, such estimates frequently overlook the comprehensive costs associated with bringing a product to market, including regulatory compliance, marketing, and overheads. While disruptive technologies have indeed led to cost-effective products, such instances are the exception rather than the norm.

Defining Innovation and Successful Innovation

Innovation transcends the mere act of inventing something that did not previously exist. It can be as nuanced as adding an incremental feature to an existing design that provides a distinct advantage. However, the litmus test for successful innovation is market acceptance—when someone finds enough value in the innovation to pay for it, transforming it into a viable commercial entity.

The crux of successful innovation is its appeal to potential customers. Young entrepreneurs, often buoyed by academic research and expertise in a particular area, may find themselves in a 'comfort zone' where they mistakenly assume that their idea will naturally attract a market. This assumption can lead to a perilous path where the product, developed in isolation from market realities, struggles to find its customer base.

Bridging the Innovation Gap

A systematic approach is vital for translating research and ideas into marketable innovations. This book aims not to dampen the enthusiasm of aspiring innovators but to lend insights that could enhance the probability of success for new ventures. It encourages entrepreneurs to conduct rigorous market research before development begins, ensuring that the innovation meets a genuine market need.

As per the insights gleaned from reviewing hundreds of startup ideas, a significant pattern emerges: the most successful innovations are not necessarily groundbreaking inventions but are often improvements or iterations that add tangible value to existing solutions. Moreover, successful innovators understand that beyond the product lies the challenge of navigating regulations, understanding market dynamics, and crafting a compelling value proposition.

In the context of the Indian market, these insights are particularly poignant. India, with its burgeoning startup ecosystem, has seen a few success stories that have harnessed the power of innovation to create impactful products. However, the journey from an idea to a successful product is riddled with challenges, including the need for thorough market understanding and the ability to anticipate and meet regulatory requirements.

The path to successful innovation is, therefore, one that is iterative and grounded in reality. It requires a willingness to step outside one's comfort zone, validate ideas against the harsh criteria of the market, and adapt based on genuine customer feedback. Innovators must embrace a holistic approach that considers not just the product but also the business model, the go-to-market strategy, and the long-term vision for their ventures.

Successful innovation is not just about the idea or the technology; it's about the value it creates for customers and the entrepreneur's ability to deliver a product that meets that value promise. This book seeks to guide innovators on this journey, helping them to navigate the complex terrain of transforming a concept into a successful product, thereby contributing to a more vibrant and impactful startup ecosystem.

Comprehensive Roadmap: Goals and Insights of This Book

Let's now delve into the heart of this book: a comprehensive roadmap that charts the course through the multifaceted terrain of product development. This is where the rubber meets the road, where theoretical concepts and innovative prototypes are honed and refined into market-ready products.

This book serves as a navigator for the journey of product development, providing detailed insights and structured strategies that transcend the basic understanding most associated with the early stages of startup growth. The aim is to offer a granular view of the product development lifecycle, drawing on the wealth of knowledge that lies beyond the foundational elements.

By aligning the innovation process with product development methodologies, entrepreneurs can ensure that their innovations are not only technically feasible and groundbreaking but also commercially viable and scalable. This holistic approach is essential in today's competitive marketplace, where successful product development is not just about creating something new but about crafting a product that resonates with customers and stands the test of time.

The Stages of Product Development: A Deep Dive

The chapters that follow will chart a course through the nuanced phases of product development, each characterized by its unique challenges and opportunities:

Ideation: The journey begins on the fertile grounds of ideation, where innovation takes its first breath. Here, we lay the foundations by harvesting the creative ideas and research insights previously discussed and shaping them into a cohesive concept aligned with market needs.

Proof of Concept: This stage is the crucible in which the idea is tested for its feasibility. It's a reality check that pits the concept against technical, commercial, and financial benchmarks to determine its practicality.

Design: With a viable concept in hand, the design phase commences. This is an iterative process of crafting and refining the product, taking into account user feedback, functionality, and aesthetics to evolve a prototype that embodies the envisioned solution.

Alpha and Beta Prototypes: These iterative prototypes represent early and more refined versions of the product, each serving as a critical step toward a market-ready offering. They are essential for internal testing (alpha) and select user testing (beta), providing invaluable feedback that informs further refinement.

Pilot Build: A bridge between prototyping and full-scale production, the pilot build is where the product is tested under production conditions. This phase helps in identifying any final adjustments needed before large-scale manufacturing begins.

Regulatory Compliances: Compliance is not an afterthought but an integral part of the development process. This stage ensures that the product adheres to all relevant industry standards and regulations, affirming its safety, quality, and market readiness.

Reliability Evaluation: Before a product can be released into the wild, it must undergo rigorous reliability testing. This phase evaluates the product's endurance, performance, and longevity under various conditions to ensure it meets the high standards expected by consumers.

Volume Manufacturing: Transitioning from limited production to mass manufacturing, this stage is where scalability is key. The focus is on maintaining the integrity of the product while optimizing the manufacturing process for cost efficiency and quality control.

Market Launch and Scaling: This critical phase involves introducing the product to the market, implementing strategies to capture consumer attention, and scaling production to meet growing demand.

Post-Launch: The product's journey doesn't end at launch. Continuous improvement is vital, driven by customer feedback, market trends, and new technological developments to ensure the product remains competitive and relevant.

As we traverse through these stages, we'll uncover the strategies that successful companies employ to ensure each phase transitions smoothly into the next. From ideation techniques that encourage out-of-the-box thinking to design philosophies that balance form and function, from compliance strategies that foresee regulatory hurdles to manufacturing practices that scale without compromise—the discussion will be rich with expert insights and best practices. Moreover, each chapter will provide a detailed exploration of these stages, replete with real-world examples, case studies, and actionable advice. Readers will not only understand the 'what' and 'how' of each phase but also the 'why'—the rationale that guides the decisions of seasoned product developers.

This comprehensive roadmap is designed to equip innovators, entrepreneurs, and product developers with the knowledge and skills necessary to navigate the complex landscape of bringing a product to market. By bridging the gap between theory and practice, this book becomes more than just a read. It evolves into a mentor for your product development journey. As you progress from one chapter to the next, let each page turn be a step forward in your entrepreneurial quest, each insight a tool in your kit, and each strategy a stepping stone towards realizing your vision.

Planting the Seeds of Innovation

In the world of product development, ideation stands as the pivotal moment where inspiration transforms into tangible reality. This chapter will dissect the unique essence of ideation in hardware development, a realm where the constraints of the physical world meet the boundless potential of the human mind. Here, ideation is not just a spark of creativity; it's the fire that fuels the journey from abstract concepts to groundbreaking, tangible innovations with the power to reshape our society.

In hardware development, ideation is a symphony of art and science. It's where abstract ideas crystallize into definitive shapes, driven by the imperative to create tangible outcomes. This critical stage is not merely about letting imagination run wild; it's about channeling creativity into practicality, molding visions into realities that can be held, used, and experienced.

Each step of ideation is a deliberate, thoughtful process, akin to a chess game where each move is calculated and strategic. It's a journey that requires more than just innovative thinking; it demands a deep dive into the practicalities of bringing a product to life in the tangible world.

The Structured Process of Hardware Ideation

Ideation in the context of hardware transcends spontaneous creativity, it's a disciplined process that demands a strategic approach:

Problem Identification: The genesis of hardware ideation lies in identifying real-world problems begging for solutions. This step is about understanding the needs and challenges that exist, setting the stage for meaningful innovation.

Solution Envisioning: Here, the focus shifts to conceptualizing potential solutions. It's a phase where creativity meets the constraints of physical design and manufacturing, blending imagination with feasibility.

Feasibility Analysis: This stage is the crucible where ideas are tested against the harsh realities of the physical world. It involves a rigorous evaluation of the proposed solutions in terms of material selection, cost implications, manufacturing processes, and market viability.

The transformative power of ideation in hardware product development is a testament to how initial sparks of creativity can lead to monumental shifts in our daily lives and societal structures. Throughout history, this process has been at the heart of some of the most significant innovations, from the inception of the wheel to the sophisticated engineering behind modern smartphones. These groundbreaking inventions all share a common origin – they began as simple ideas, which through meticulous and strategic development, blossomed into tangible realities with far-reaching impacts.

Revolutionizing Industries and Shaping Interactions

The journey of ideation in hardware development is not just about creating new products; it's about reimagining and reshaping entire industries. It's about altering the very manner in which we interact with the world around us. Consider the evolution of renewable energy technologies like solar panels and wind turbines. These innovations originated from the ideation process aimed at tackling one of the most pressing global challenges: sustainable energy production. They are not merely products; they represent a fundamental shift in how humanity harnesses and interacts with energy resources.

Balancing Imagination with Practicality

At its core, effective ideation in hardware development is a balance between imagination and practicality. It requires an in-depth understanding of the physical world's constraints, combined with technical knowledge and a sensitivity to societal and environmental needs. This balanced approach ensures that the ideation process, while ambitious and innovative, is firmly rooted in feasibility and applicability.

Take, for example, the evolution of electric vehicles (EVs). The concept of EVs emerged from a blend of environmental consciousness and technological advancements in battery and automotive design. This innovation encapsulates a deeper understanding of the need for sustainable transportation solutions – solutions that align with societal needs and environmental preservation.

The development of EVs is a prime illustration of how ideation, fueled by a blend of technological insight and environmental concern, can lead to a product that not only offers a new way of transportation but also contributes to a broader societal shift towards sustainability.

Bringing Social Impact to the Forefront of Ideation

Identifying Societal Needs

The process of identifying societal needs requires a deeper dive than mere surface-level observation. It demands an intimate understanding of the complex nature of societal challenges:

Understanding Multifaceted Issues: For instance, tackling poverty isn't just about financial assistance; it's about understanding its impact on education, health, and economic stability. In addressing issues like water scarcity, it's crucial to comprehend the underlying societal and infrastructural factors contributing to these challenges.

Community Engagement and Empathetic Design: Entrepreneurs must immerse themselves in the communities they aim to serve, adopting a design thinking approach that is rooted in empathy. This involves listening to the needs and experiences of community members, collaborating with social workers, economists, and environmentalists to gain a holistic view of the problems at hand.

Technological Innovation Meets Societal Needs: A perfect example of this approach can be seen in the development of affordable solar-powered lighting solutions in regions with limited access to electricity. These innovations, born from a deep understanding of specific community needs, not only provide a technological solution but also profoundly impact the daily lives and economic prospects of these communities.

Consider the case of life-saving medical devices designed for low-resource settings. Innovations like portable, solar-powered ultrasound machines or low-cost incubators for premature infants have been game-changers in regions where traditional medical infrastructure is lacking. These devices were developed with a keen awareness of the specific challenges and limitations of these environments, demonstrating how hardware innovation, guided by a nuanced understanding of societal needs, can have a profound and life-changing impact.

In the pursuit of meaningful hardware innovations, understanding and bridging societal divides is not just a noble goal, but a crucial step in the ideation process. Entrepreneurs, especially in the field of hardware product development, must delve deep into the realities of those facing daily challenges, transcending beyond mere observation to actual experience. It's about empathetically understanding the barriers – physical, financial, educational – that separate technology's potential from its actual beneficiaries.

Understanding and Bridging the Divide

The process of assessing the divide requires more than superficial understanding. Entrepreneurs must immerse themselves in the environments of their intended users, identifying barriers that impede access and usage. For instance, in the agricultural sector, the introduction of affordable, robust farming equipment has revolutionized the lives of smallholder farmers, enhancing their productivity and economic stability. Similarly, in education, digital platforms that are accessible and user-friendly have democratized learning, breaking down barriers imposed by geography and socio-economic status.

This phase involves a critical analysis of market failures and social inequities. Human-Centered Design (HCD) becomes an invaluable tool here, enabling the development of solutions that are not only technologically sound but also empathetic to the users' needs and conditions.

Leveraging Technology Readiness Levels for Social Innovation

Technology Readiness Levels (TRL) provide a structured approach to assessing the maturity of a technology, from the conceptual stage to its market readiness. This metric is particularly valuable in social innovation, guiding entrepreneurs through the nuanced process of technology development.

For example, consider the space of water purification technologies. In regions plagued by waterborne diseases, a technology with a high TRL signifies that it's not just a concept but ready for real-world deployment. Such innovations can drastically alter the public health landscape of these areas.

The TRL framework aids in systematically planning the development process, identifying potential challenges and ensuring the robustness of the technology for diverse social settings. For instance, the development of portable, solar-powered water purifiers has significantly impacted communities lacking access to clean water. These devices, having progressed through various TRL stages, are now providing a sustainable solution for clean drinking water, profoundly impacting public health and community resilience.

Designing for Inclusivity in Hardware Product Development

Inclusive design emerges as a cornerstone in hardware product development, ensuring that products cater to a wide and diverse user base. This approach to design recognizes and embraces human diversity in all its forms – ability, language, culture, gender, and age. It's about creating products that are not just universally usable but also resonate with a broad spectrum of the human experience.

Embracing the Full Spectrum of User Diversity

Inclusive design in hardware necessitates a deep understanding of the user's world. It's about acknowledging that a one-size-fits-all approach is insufficient. For example, consider the design of user interfaces in health-monitoring devices. By making these interfaces intuitive for both tech-savvy youths and the elderly, developers can significantly broaden the device's appeal and usability. This approach not only increases adoption rates but also ensures that the benefits of such technology reach all segments of society.

Moreover, inclusive design demands empathy. It requires products to be physically accessible, culturally sensitive, and economically affordable. This principle is particularly vital when developing products for marginalized or underserved communities, where accessibility and affordability can be significant barriers.

Contemplating the Long-Term Impact

Looking beyond immediate solutions, entrepreneurs engaged in hardware development must consider the long-term societal impact of their products. This foresight involves a holistic view of sustainability – encompassing environmental, economic, and social dimensions.

For instance, in the development of renewable energy solutions like solar panels or wind turbines, it's crucial to assess not only their environmental benefits but also their socio-economic impact. This includes factors like job creation in the renewable energy sector and the broader implications for community development and climate change mitigation.

Another notable instance is the development of low-cost prosthetics. These devices, designed with both affordability and functionality in mind, have transformed the lives of individuals in low-income countries, offering them mobility and independence. An additional example is the design of energy-efficient, durable household appliances targeted at lower-income households. These products, while economically accessible, help reduce energy consumption and lower utility bills, contributing to both environmental sustainability and economic resilience for these families.

Agriculture & Rural Development: Catalyzing Change through Innovation

In the agricultural sector, especially within rural landscapes, innovation can be a significant game-changer. Small-scale farmers, who often form the backbone of rural economies, stand to benefit immensely from technological advancements. The introduction of mechanized tools specifically designed for smaller plots can revolutionize their work, significantly increasing productivity while reducing the physical toll of labor-intensive practices.

Empowering Small-Scale Farmers through Technology

One area ripe for innovation is the development of affordable, solar-powered machinery. Such technology is not only environmentally sustainable but also perfectly suited for areas where access to conventional power sources is limited. Furthermore, digital platforms that connect farmers directly to broader markets can disrupt traditional supply chains that often marginalize these small-scale producers. By leveraging these platforms, farmers gain fairer prices and better access to markets, enhancing their income and economic stability.

Precision agriculture is another frontier, harnessing the power of data analytics to optimize resource usage and maximize crop yields. This technology enables farmers to make informed decisions about planting, irrigation, and harvesting, thus enhancing efficiency and productivity.

A tangible example in agricultural innovation is Hello Tractor, an "Uber for tractors" service in Africa. It provides farmers with access to tractor services on demand, improving their productivity and efficiency.

Water & Environment: Solutions for Sustainability

Addressing water scarcity and pollution requires innovative solutions that are both effective and sustainable. Low-cost water purification systems powered by renewable energy sources can provide safe drinking water in areas affected by pollution and scarcity. These systems are particularly crucial in regions where waterborne diseases are prevalent, and access to clean water is a critical challenge.

Smart irrigation technologies present another opportunity for innovation. By reducing water wastage, these systems ensure that water resources are used efficiently, which is vital in areas where water is scarce. Additionally, bioremediation techniques offer a natural and effective way to clean pollutants from ecosystems. Using plants or microbes to detoxify areas affected by industrial waste not only restores these environments but also preserves biodiversity.

Products like LifeStraw have revolutionized access to clean drinking water, using simple, portable filtration systems suitable for areas with limited resources.

Financial Inclusion: Bridging the Gap with Innovative Solutions

In the financial sector, the focus shifts to designing products that cater to the unbanked population. Innovative solutions like mobile-based banking platforms are crucial in areas devoid of traditional banking infrastructure. These solutions provide essential financial services, from savings to transactions, right at the fingertips of users in remote areas.

Microfinance services are another area of innovation, providing small-scale entrepreneurs with the credit needed to grow their businesses. This financial support can be a critical factor in kickstarting local economies and fostering entrepreneurship.

Tailored insurance products also play a vital role in creating financial resilience, especially for populations in rural areas who may face unique risks and challenges. By designing insurance solutions that cater to the specific needs of these communities, entrepreneurs can protect them against unforeseen financial shocks.

Women's Safety: Harnessing Technology for Empowerment

In today's world, where women's safety remains a pressing concern, technology offers innovative solutions that can provide both protection and empowerment. Advancements in wearable technology have led to the development of discreet devices capable of sending emergency alerts with the touch of a button. An exemplary case is the development of smart jewelry, like the Nimb Ring, which allows the wearer to send a distress signal to pre-selected contacts, including location details, with a simple press.

Furthermore, mobile applications like Safetipin, which use community-reported data to map safer routes in cities, have empowered women to make informed decisions about their travel. These apps, by crowdsourcing data on well-lit streets, populated areas, and safe public spaces, provide a layer of security and awareness.

Health: Bridging the Accessibility Gap

In the health sector, the need for affordable solutions in rural areas is critical. One groundbreaking innovation in this realm is the development of portable diagnostic tools, such as the Butterfly iQ, a handheld ultrasound device. This technology, which can connect to a smartphone, brings diagnostic capabilities to remote areas where traditional medical imaging equipment is unavailable.

Telemedicine services are another transformative innovation. Platforms like TeleDoc have enabled patients in rural areas to consult with specialists, overcoming geographical barriers. Additionally, innovations in drug delivery systems, such as heat-stable vaccines and medications that do not require refrigeration, are crucial for managing chronic diseases in regions with limited access to healthcare infrastructure.

Education: Technology as a Bridge

In education, technology has the potential to revolutionize learning, especially in underserved communities. E-learning platforms, tailored with localized content, can make education more accessible and relevant. For instance, Khan Academy's vast range of educational resources, available in multiple languages, has been instrumental in providing quality education worldwide.

Mobile applications designed for interactive learning experiences have also shown significant impact. Apps like Duolingo, with its engaging and user-friendly interface, have made language learning accessible to a broader audience. Additionally, technology-enabled teacher training programs can enhance the quality of education, equipping educators with modern teaching tools and methodologies.

Innovations that support lifelong learning and skill development are also crucial in today's dynamic job market. Platforms like Coursera and LinkedIn Learning offer courses that cater to ongoing education and skill enhancement, aligning with the evolving professional landscape.

Clean Energy: Empowering Communities at the Grassroots

In the pursuit of clean energy solutions, the emphasis on efficiency and affordability is crucial, particularly for driving adoption at the grassroots level. Small-scale, decentralized energy systems like off-grid solar installations have begun to revolutionize power accessibility in remote areas. A notable example is the work of companies like M-KOPA Solar, which provides affordable solar-powered lighting and charging systems to households in East Africa. These systems not only offer clean energy but also eliminate the dependence on expensive and polluting kerosene lamps.

Community wind turbines represent another sustainable energy solution, offering a localized approach to harnessing wind energy. In Scotland, for instance, community-owned wind projects have been successful in providing power and generating revenue for local development.

Biogas systems, which convert agricultural waste into energy, are also gaining traction. In countries like India, biogas plants are being used not just for energy production but also for improving soil fertility through the byproduct – bio-slurry.

Livelihoods: Enhancing Economic Opportunities for Rural Communities

In rural areas, where traditional livelihoods are often under threat, access to innovative tools and platforms can make a significant difference. E-commerce platforms that bridge the gap between local artisans and global markets have been transformative. Websites like Etsy have allowed artisans to reach a broader audience, increasing their income and preserving traditional crafts.

Additionally, tools and technologies that enable efficient and sustainable practices in crafts and agriculture can significantly impact livelihoods. For example, innovations in textile machinery that are energy-efficient and less water-intensive are enabling a more sustainable production process.

Cooperatives play a crucial role in pooling resources and enhancing bargaining power for small producers and artisans. An example is the Amul cooperative in India, which revolutionized the dairy industry by empowering local farmers.

Physical Disability

For individuals with disabilities, assistive technologies can dramatically improve their quality of life. Innovations may include affordable prosthetics, mobility aids, adaptive interfaces for computers and mobile devices, and accessible design in public infrastructure. These technologies not only empower individuals with disabilities but also enable their fuller participation in society.

Building upon the foundation of impactful innovation, let's integrate a case study that epitomizes the seamless blend of entrepreneurship, technological sophistication, and societal benefit—the development of a non-invasive oral cancer detection system.

Case Study: Non-invasive Oral Cancer Detection System

Oral cancer, a prevalent and often fatal disease, is particularly insidious due to its late detection. Recognizing the dire need for early diagnosis, a team of entrepreneurs and biomedical engineers turned their attention to the development of a non-invasive detection system. This innovative device employs cutting-edge biophotonics and imaging technology to identify cancerous lesions at an early stage, all without the need for painful biopsies or the anxiety-inducing wait for results. The below diagrams in sequence show how a concept was translated to a product.

Need identification

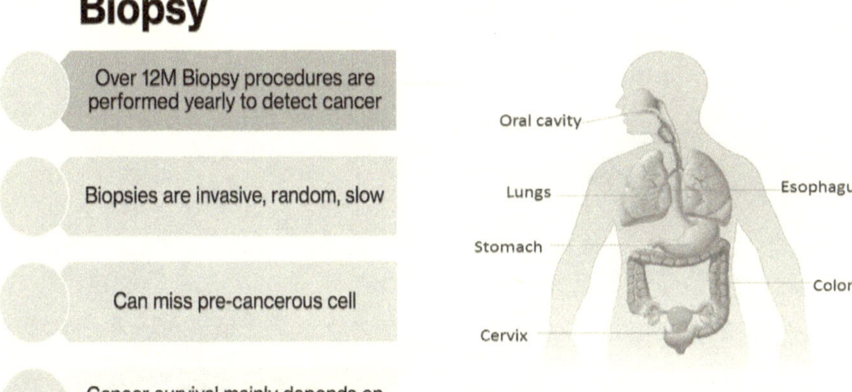

Conventional vs Optical Biopsy

❖Doctor will examine, do incisions and take out tissues from suspicious sites and send for histo-pathological examination

- Painful, discomfort to patients
- Multiple biopsies
- Can miss precancerous cells
- Takes days to get results

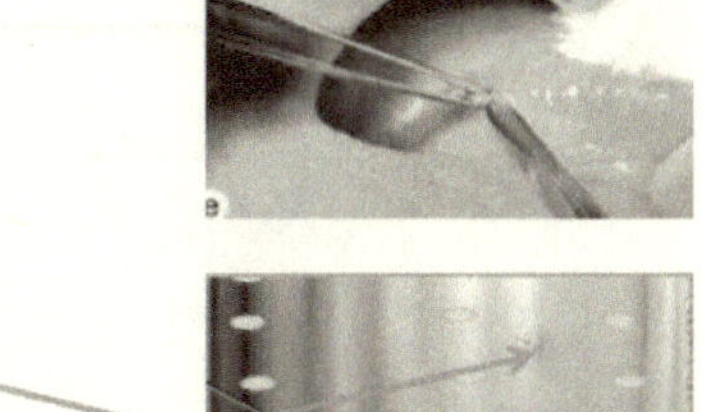

Objective is this to come up with a Non invasive biopsy device which uses light and instantaneously detects cancer specifically early cancer

Proof of Concept in the laboratory

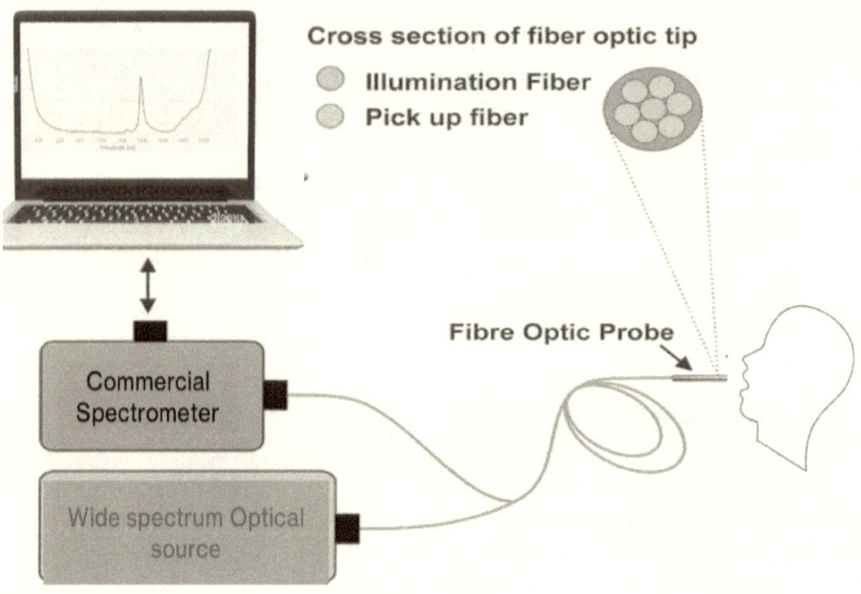

Product Architecture and System Design

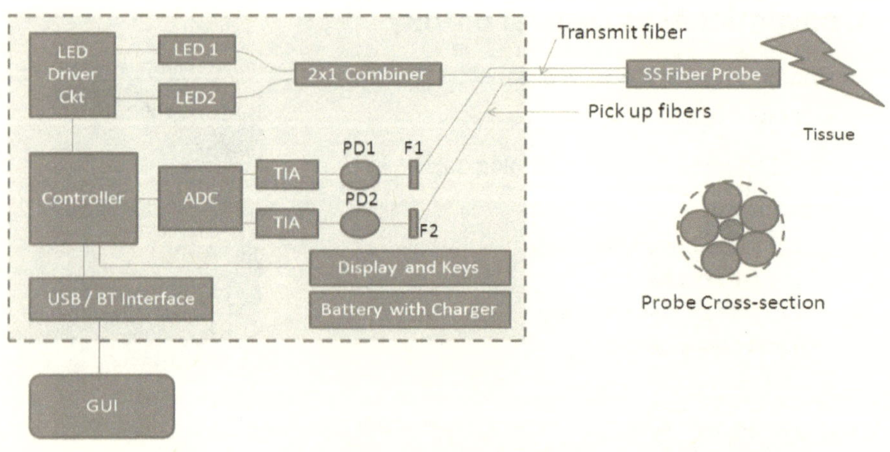

Translating to Design for Manufacturability

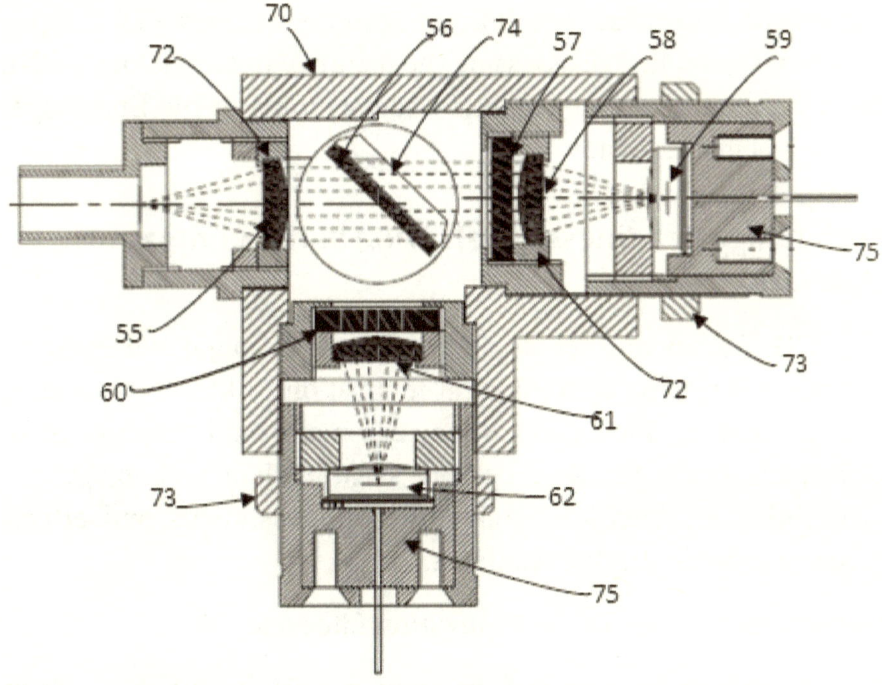

The Alpha Prototype

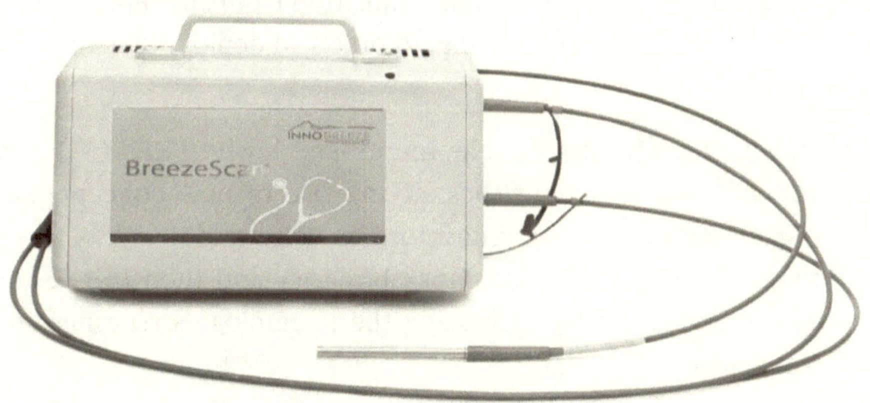

As we continue to unravel the multifaceted process of innovation in hardware product development, we arrive at a crucial evaluative framework known as Technology Readiness Levels (TRL). This scale is instrumental in assessing the maturity of technology—from the nascent stages of conceptual understanding to the final stages where it is market-ready.

Technology Readiness Levels (TRL) in the Context of Innovation

The TRL framework serves as a navigator through the complex journey of bringing technological innovation to life. By providing a clear metric, TRL enables entrepreneurs and innovators to objectively assess the development stage of their technology and make informed decisions about further investments and efforts required to reach market readiness.

Understanding TRL in the Innovation Lifecycle

Basic Research to Market Introduction: TRL begins at 1, where the basic principles of the technology are observed, and ends at 9, where the technology has been proven through successful practical use. This journey encapsulates the transition from conceptual ideas, through development and testing, to practical deployment and user acceptance.

Guiding Research and Development: The lower TRLs (1-3) are where basic research flourishes, giving rise to new concepts and early-stage prototypes. As innovators move to middle TRLs (4-6), the focus shifts to validation in a laboratory and then in relevant environments, gradually increasing the technology's readiness for the real world.

From Prototypes to Products: The higher TRLs (7-9) are where prototypes become products. System demonstration in operational environments (TRL 7) is followed by the completion and qualification of the actual system (TRL 8), culminating in the technology's successful application in its final form (TRL 9).

Applying TRL to Hardware Innovation

In hardware innovation, TRL helps in making critical decisions at each stage. For instance, at TRL 4 or 5, an innovator might decide to pivot based on feedback or technical challenges encountered during validation. By TRL 7, they should be engaging with potential users to test the system in the field, and by TRL 9, they should be ready to scale up production and distribution.

TRL and Social Impact

Applying the TRL framework to socially impactful hardware innovations require a nuanced approach. Innovators must consider how the technology will not only function but also how it will be received by society. For example, an innovation that reaches TRL 9 in clean energy must not only be technically sound but also culturally acceptable and economically viable for the communities it aims to serve.

The TRL framework is more than a checklist; it is a strategic tool that provides a structured pathway for the development of hardware innovations. It helps ensure that products are not only technologically advanced but also socially relevant and ready for the challenges of the real world. As innovators navigate through the TRLs, they do so with the knowledge that each level brings them closer to delivering solutions that can make a tangible difference in society.

Self-Check on Innovative Ideas:

Originality of the Idea: The Bedrock of Innovation

In the journey of hardware product development, the initial phase of validating the originality of an idea is a critical step. This process is not just about ensuring novelty but also about establishing a firm belief in the idea's potential to bring something unique and valuable to the market.

The originality of an idea is the cornerstone of any successful hardware innovation. Whether it emerges as a sudden inspiration or as an evolution of existing concepts, the key lies in the entrepreneur's conviction in its uniqueness. This belief is more than just a gut feeling; it is an informed understanding that the idea brings something new or significantly improved to the table.

For instance, consider the development of Dyson's vacuum cleaners. The idea of using cyclonic separation to create a bagless vacuum cleaner was a significant departure from existing products, demonstrating how adaptation and improvement of existing concepts can lead to groundbreaking innovations.

Assessing Originality through Research and Introspection

The process of establishing originality requires entrepreneurs to engage in thorough research and introspection. This involves a critical assessment of whether the idea offers a new solution to a problem or significantly improves upon existing ones. It's not just about being different; it's about adding value in a meaningful way.

Prior Art Status: Understanding the Innovation Landscape

A comprehensive understanding of existing inventions, patents, and the broader technological landscape is essential. This step ensures that the new idea does not inadvertently infringe on existing intellectual property. Engaging with IP attorneys and patent specialists is crucial in this phase. They provide expertise in identifying prior art and assist in distinguishing the new idea with unique claims and functionalities.

A real-life example of this process can be seen in the development of Tesla's electric vehicles. While electric cars were not a new concept, Tesla's approach to battery technology, vehicle design, and software integration was distinctive enough to set it apart from previous attempts, demonstrating an understanding of prior art while pushing the boundaries of technology.

This exercise is not just a legal safeguard but also a strategic exploration of where the idea fits within the current state of technology and how it pushes the boundaries of what is already known.

Patentability: Navigating the Criteria for Hardware Innovation Protection

Securing a patent is a critical step that serves as a protective shield for an entrepreneur's innovation. The journey to obtaining a patent is navigated through several key criteria that assess the viability and uniqueness of the idea.

Novelty: The invention must introduce a new concept or application that has not been publicly disclosed before. It's about breaking new ground, not rehashing known ideas. For instance, the invention of Gorilla Glass by Corning Incorporated was a breakthrough in durable, scratch-resistant glass, a novel creation at the time that significantly differed from existing glass technologies.

Non-Obviousness: This criterion ensures that the invention is not an obvious extension of existing products. It must represent a significant leap forward, not just a minor improvement or modification. A real-life example is the transition from incandescent light bulbs to LED technology, which represented a significant advancement in energy efficiency and longevity.

Industrial Applicability: The invention must have practical utility and be applicable in an industry. This means that the idea should be more than a theoretical concept; it should have a clear use in a practical setting. For example, the development of 3D printing technology had a profound impact across various industries, from manufacturing to healthcare, due to its wide range of practical applications.

Market Analysis: Understanding the Economic Landscape

Conducting a thorough market analysis is crucial for determining the potential success of a hardware product. This involves several key components:

Market Trends and Customer Segments: Entrepreneurs need to understand current market dynamics and identify potential customer bases. For example, the rise of wearable technology like fitness trackers has been driven by growing health consciousness among consumers.

Competitive Analysis: This involves examining existing products in the market, identifying their strengths and weaknesses. A classic case is the competition in the smartphone industry, where companies continuously analyze each other's products to innovate and capture market share.

Product Positioning and Strategy: Insights gained from market and competitive analysis inform decisions about product positioning, pricing, and promotional strategies. An example is Tesla's positioning of its electric vehicles not just as environmentally friendly alternatives but as high-performance, luxury cars.

Unique Selling Proposition (USP): Crafting a Distinct Identity in Hardware Innovation

In the competitive landscape of hardware product development, the Unique Selling Proposition (USP) is the linchpin that can set a product apart from its competitors. It's that distinctive feature or benefit that not only resonates with customers but also defines the product's identity in the market. A USP could be an innovation in efficiency, a novel feature that expands the product's utility, or superior performance metrics that eclipse those of competitors.

For instance, the development of the GoPro camera exemplifies a potent USP. Its compact size, robust design, and ability to capture high-definition videos in extreme conditions made it a favorite among adventure enthusiasts, clearly distinguishing it from other portable cameras on the market.

Creating a USP involves a blend of ingenuity, market awareness, and an acute understanding of consumer needs and preferences. It's about crafting a value promise that is not only compelling but also tangible and deliverable.

Development Cost and Time: Navigating the Practicalities of Bringing a Product to Market

The journey from concept to market in hardware development is rife with practical challenges, particularly concerning development costs and timeframes. A realistic assessment of these factors is crucial to avoid pitfalls like budget overruns and project delays.

Accurately estimating development costs encompasses a range of expenses, including materials, labor, research and development, and potential marketing and distribution costs. Timeframe estimations must account for each development stage, from initial design and prototyping to testing, production, and market launch.

A case in point is the development of Dyson's Air Multiplier, the bladeless fan. The R&D process, which involved extensive prototyping and testing, was both time-intensive and costly. However, the company's meticulous planning and budgeting ensured the successful launch and market penetration of this innovative product.

Strategic Considerations in USP and Cost-Time Analysis

Defining a USP and assessing development costs and timeframes are not isolated tasks; they are interlinked and critical to the product's overall success. A well-defined USP can justify higher development costs if it leads to a unique product that captures a significant market share. Conversely, a realistic appraisal of costs and development timelines ensures that the project remains feasible and sustainable, even with an ambitious USP.

Product Cost: Striking a Balance Between Value and Viability

In hardware product development, determining the right product cost is a critical exercise that directly influences market success. The cost must reflect the value offered while remaining competitive and attractive to consumers. It's a delicate equilibrium between covering the production, distribution, and marketing expenses and ensuring a profitable return, all while being mindful of customer price sensitivity.

For example, the launch of the Raspberry Pi computer showcased how striking the right balance between cost and functionality could lead to market success. Priced affordably, it offered substantial computing power, appealing to a wide range of users from hobbyists to educators, thus capturing a significant market share.

Repeatability and Manufacturability: Ensuring Consistent Quality at Scale

A product's success in the market hinges not only on its initial quality but also on its ability to maintain this quality consistently across all units. Repeatability in manufacturing is vital for building consumer trust and upholding the brand's reputation. This means that the selection of materials, manufacturing processes, and quality control measures must be meticulously planned to ensure uniformity in every product.

Achieving high levels of manufacturability and repeatability might require substantial investments in advanced manufacturing technologies and process optimization. A prime example is the production of Apple's iPhone. The company's investment in precision manufacturing techniques has enabled it to produce millions of units while maintaining stringent quality standards.

Sustainability: A Cornerstone of Modern Product Development

Sustainability is no longer an optional consideration in product development; it's a crucial component of a product's appeal and longevity. This encompasses environmental considerations, such as the use of eco-friendly materials and energy-efficient production processes, as well as the product's adaptability to changing market trends and resources.

A notable example in this area is Tesla's electric cars. The company's focus on sustainable energy and eco-friendly manufacturing processes has not only reduced environmental impact but also aligned with consumer trends favoring green technology.

Leveraging Business Models for Comprehensive Analysis

The ideation phase, enriched with these insights, is crucial for developing products that are not only innovative but also market-ready and legally sound. Utilizing tools like the Stanford Business Canvas Model can offer a comprehensive view of the business model, helping entrepreneurs to consider all aspects necessary for turning their ideas into successful products.

Overcoming Obstacles in Hardware Development and Mastering Project Management

Consider the case of Juicero, a startup that became infamous in the hardware industry for its high-end, internet-connected juicing machine. The company had a compelling vision: to revolutionize home juicing with a machine that could press proprietary packets of pre-chopped fruits and vegetables, delivering convenience and fresh juice with the press of a button.

From its outset, Juicero's journey was fraught with hardware development delays. The initial design, while sleek and innovative, proved complex and costly to manufacture. The intricacies of designing a machine capable of exerting four tons of force—enough to squeeze juice efficiently from a packet—led to numerous prototypes, each iteration trying to balance functionality with manufacturability and cost. These design challenges resulted in significant delays, pushing back the product launch and straining investor patience.

Moreover, the startup's ambition to integrate advanced technology into its juicer, such as QR code scanning and Wi-Fi connectivity for tracking the freshness of produce packets, further complicated the development process. Each added feature meant more testing, more potential points of failure, and more time before the product could be brought to market.

The impact of these delays was multifaceted. Not only did they escalate the cost of the juicer, making it less accessible to the average consumer, but they also eroded the window of opportunity Juicero had to dominate the market. By the time the product was ready, competitors had taken notice and begun exploring simpler, more cost-effective solutions. The market advantage that Juicero might have had with a timely launch was severely diminished.

Juicero's story is a cautionary tale in the hardware startup world, highlighting how delays can compound, affecting not just the bottom line but also the brand's reputation and market position. It speaks to the necessity of a balanced approach to innovation— one that is ambitious yet tempered with practical considerations of manufacturability, cost, and the patience of both investors and the marketplace.

Hardware development delays, as exemplified by Juicero's experience, underscore the delicate interplay between innovation, product development, and market dynamics. They serve as a potent reminder to hardware entrepreneurs that while meticulous planning and innovative design are crucial, they must be weighed against the realities of bringing a complex hardware product to market in a timely and cost-effective manner. This balance is what ultimately enables a hardware product to not just enter the market but to capture and maintain a foothold in it.

Analyzing the Impact: The Ripple Effects of Delays in Hardware Development

Time dictates the rhythm in hardware development, and delays can disrupt the entire framework. The repercussions of these delays resonate through every facet of a startup's journey, influencing not only the immediate path to market but also the broader lifecycle of the enterprise.

The Competitive Edge in Jeopardy

The impact of delays on market competitiveness is immediate and often severe. In the hardware domain, where product lifecycles are intricately tied to technological advancements and consumer trends, a delay can mean ceding ground to competitors who are quicker to market. The loss of first-mover advantage is particularly detrimental in sectors where consumers quickly align with brands that are seen as pioneers. Delayed market entry can relegate a startup's offering to the status of a follower rather than a leader, altering consumer perception and potential market share.

Funding and Financial Viability

From a financial standpoint, delays can strain the delicate lifelines of capital that sustain a startup. Investors anticipate returns on a timeline, and when progress stalls, it can lead to a tightening of the purse strings, or worse, a complete withdrawal of financial support. For a startup, this can mean a scramble to manage operational costs without the anticipated revenue from product sales, leading to a potential cash flow crisis.

Brand Reputation and Customer Trust

The consequences of delays extend to the intangible yet invaluable currency of brand reputation. In the age of instant communication, a startup's failure to deliver on time can quickly escalate into public relations challenges. Customers who face delays might express dissatisfaction across social platforms, leading to a broader erosion of trust in the brand. In the long-term, this can dampen enthusiasm for future product releases and tarnish the company's image.

Operational Morale and Team Dynamics

Within the operational domain of the startup, the effects of delays percolate through the morale and dynamics of the team. Persistent delays can lead to frustration, a dampening of the innovative spirit, and even the loss of key personnel who may seek more stable or rewarding opportunities elsewhere. For a startup, where each team member's contribution is magnified, this can lead to significant disruptions in the development process.

Market Dynamics and Customer Expectations

In parallel, the market does not stand still. Delays can mean that a product, once perfectly attuned to customer needs and expectations, becomes less relevant or even obsolete. This misalignment with market dynamics can necessitate further development and additional features, leading to a vicious cycle of delays.

Adaptation and Evolution

Finally, delays can have a profound impact on the startup's lifecycle, affecting its ability to adapt, evolve, and survive. A delay in one product can impact the development timelines of future products, disrupt long-term strategic planning, and stifle the growth trajectory of the company. For a startup, the ability to learn from delays and build a more resilient development process can be the difference between flourishing and floundering.

Normally startup forms a team with their friends and colleagues. As the product release gets delayed, there would be parental pressure to these youngsters to leave this initiative and take up some corporate job. Finally, there is possibility of team getting dismantled/diluted.

Another point: Team selection. It is not that close friends become the founders and start taking up responsibilities like CEO, CTO, CFO etc. The team shall be formed with mutually complementing expertise/strengths.

Identifying Core Causes of Delays

The Criticality of Design for Manufacturability (DFM)

DFM is not merely a set of guidelines—it is a philosophy that must permeate every aspect of hardware design and development. Its influence on the timeliness and success of a project cannot be overstated. By elevating DFM to a position of primary importance, hardware developers and startups can mitigate one of the core causes of development delays, maintaining momentum and ensuring that their product reaches the market poised for success. This principle is the guardian of efficiency and the gatekeeper of smooth transitions from design to production. Neglecting DFM equals building a bridge without considering the strength of the river's current—a recipe for structural collapse.

The Central Pillar of DFM in Hardware Development

DFM is the discipline that ensures hardware is designed with the manufacturing process in mind. It is a holistic approach that takes into account every aspect of production, from the selection of materials and components to the methods and environments of manufacturing. DFM is not a single checkpoint but a continuous thread that runs through the entire development process.

The Multifaceted Implications of Ignoring DFM

Manufacturing Process: At the heart of DFM lies the selection of an appropriate manufacturing process. Ignoring DFM can lead to designs that are not aligned with the most efficient production methods, resulting in higher costs and longer lead times.

Product and Part Design: Every curve, edge, and angle of a hardware component must be designed for manufacturability. Overlooking DFM can result in designs that are difficult or impossible to manufacture with standard tooling or techniques, necessitating costly and time-consuming custom solutions.

Material Selection: The choice of materials impacts not only the cost and durability of the product but also its manufacturability. Inappropriate material choices can lead to production delays as manufacturers struggle with unexpected challenges like excessive wear on tooling or difficulties in assembly.

Environmental Considerations: The manufacturing environment, including temperature, humidity, and cleanliness, can have a profound effect on production yield and quality. Designs that do not take these factors into account may suffer from high rates of defect or failure, leading to delays as issues are diagnosed and rectified.

Compliance and Testing: Hardware products must often meet stringent regulatory and industry standards. Designs that fail to incorporate necessary compliance features from the outset can face significant delays during the testing and certification phases.

The Ripple Effect of DFM Neglect: When DFM is not integrated into the design process, the repercussions ripple outwards, affecting not only the initial production but also aspects like ease of assembly, serviceability, and end-of-life disposal.

Ignoring DFM can stall even the most promising hardware projects. It can transform what should be a routine production run into a troubleshooting marathon. Conversely, a robust DFM approach can streamline production, reduce costs, and ensure that the product moves from the prototype stage to the hands of consumers without unnecessary delays.

Certification Test Failures: Navigating the Compliance Landscape

Certification is the rigorous test of a hardware product's adherence to industry and regulatory standards, and its importance in the product development lifecycle cannot be overstated. Certification test failures can be major stumbling blocks, often unforeseen by even the most experienced hardware developers, and the implications are both costly and time-consuming.

The High Stakes of Certification in Hardware Development

Certification is not a single hurdle but a series of barriers that a product must clear to be deemed safe, reliable, and compliant. The stakes are incredibly high, as certification validates a product for market entry. Failure to pass these tests can mean going back to the drawing board, incurring additional R&D costs, and pushing back launch dates—each contributing to budget overruns and missed market opportunities.

The Intricacies of Certification Compliance

A common pitfall in the certification process is underestimating its complexity. There's a myriad of certifications applicable to hardware products, including but not limited to electromagnetic interference (EMI) and electromagnetic compatibility (EMC) standards. Each certification has its own set of requirements, tests, and standards that vary not only by industry but also by country.

The Cost of Misunderstanding Certification Requirements

A misunderstanding or lack of awareness of the necessary certifications can be particularly detrimental. For example, a product intended for international markets may pass domestic certification but fail to meet the standards of another country. This not only delays the product's entry into that market but also may necessitate a redesign to comply with international regulations, adding further to the cost and extending the time to market.

International Certification Standards

The complexity of certification is magnified when dealing with international markets. Each region has its own regulatory bodies and standards, and compliance in one market does not guarantee compliance in another. The European Union's CE marking, the

United States' FCC certification, and China's CCC mark are just a few examples of the diverse and intricate regulatory landscape that a hardware product must navigate.

The Domino Effect of Certification Delays

Certification delays can set off a domino effect that touches every aspect of a startup's operations. From investor relations to supply chain logistics, a delay in certification can have far-reaching consequences. It can alter production schedules, impact marketing campaigns, and disrupt the entire go-to-market strategy.

Proactive Strategies for Certification Success

Proactive strategies are essential for avoiding certification failures. This involves engaging with certification experts early in the design process, building prototypes that consider certification requirements, and staying abreast of the latest regulatory changes. Rigorous pre-compliance testing can also uncover potential issues before they result in actual certification failures.

A comprehensive understanding of certification requirements and a proactive approach to compliance are indispensable for any hardware startup. By prioritizing certification readiness from the outset, developers can avoid costly setbacks, ensuring that their product meets all necessary standards and reaches the market in a timely and cost-effective manner.

Reactive Design Changes: Confronting Testing Failures in Hardware Development

In hardware development, the transition from design to testing is a critical phase where the theoretical meets the tangible. It is at this juncture that many projects encounter a pivotal challenge: reactive design changes resulting from testing failures. These changes, often necessary to rectify defects and meet operational standards, can lead to significant project delays and disruptions.

The Disconnect Between Design and Practical Functionality

Hardware development is frequently a tale of two narratives: what appears flawless on paper may unravel in the testing phase. This disconnect arises because theoretical designs do not always account for the complexities and unpredictability of real-world conditions. Components that function perfectly in isolation might falter when integrated into a complex hardware system. Similarly, environmental factors such as temperature, humidity, component and fabrication tolerances and usage patterns can expose unforeseen flaws in the product.

Identifying Defects Post-Testing

The testing phase is the first real encounter of the product with practical scenarios. It is during this phase that many defects come to light. These may range from minor issues that require simple tweaks to major design flaws that necessitate a complete overhaul of the product. The latter can be particularly detrimental, as they require going back to the drawing board, significantly delaying the development process.

The Time-Intensive Nature of Addressing Design Flaws

Addressing these design flaws is not a trivial matter. It involves a thorough analysis to pinpoint the root cause of the failure, followed by a series of decisions on how best to rectify the issue. This may involve changing components, which can have a cascading effect on the entire design, including size, weight, power consumption, and even user interface. Each change must be carefully considered to maintain the integrity and usability of the final product.

Trade-offs and Decision-Making

One of the critical aspects of addressing design changes is the trade-offs that must be made. For instance, rectifying a flaw might mean downgrading certain specifications or opting for more expensive components, both of which can impact the final product's price, performance, and appeal. These decisions require a delicate balance between technical feasibility, cost implications, and market expectations.

Project Delays and Escalating Costs

Reactive design changes inevitably lead to project delays. Redesigning components, retesting the product, and ensuring compliance with regulatory standards all add time to the development cycle. These delays can have ripple effects, impacting launch schedules, marketing plans, and financial projections. Additionally, the cost of redesigning and retesting can strain the project's budget, further complicating the development trajectory.

Mitigating Reactive Design Changes

To mitigate the impact of reactive design changes, it is crucial to adopt a proactive approach in the design phase. This involves incorporating robust testing protocols early in the development process, engaging in iterative design practices, and building flexibility into the design to accommodate potential changes. Additionally, involving cross-functional teams in the design process can help anticipate potential issues, ensuring that the product is robust and ready for the rigors of real-world use.

Reactive design changes are an inevitable reality in hardware development, but their impact can be managed through strategic planning, early testing, and a flexible approach to design. By understanding the potential for testing failures and preparing for the subsequent design challenges, hardware developers can navigate these hurdles with minimal disruption, keeping their projects on track towards successful market entry.

Specification Evolution: Managing Changes in Hardware Development

In hardware development, the evolution of specifications is a common yet challenging phenomenon. Changes in specifications, whether they stem from customer feedback, technological advancements, or shifts in market demands, have far-reaching implications on the product development timeline.

The Nature of Specification Changes

Specification changes in hardware development are often inevitable. As the product progresses from concept to prototype to final design, new insights, feedback, and ideas emerge. This feedback loop is invaluable, as it ensures that the final product is well-tuned to customer needs and market trends. However, these changes, even when minor, can set off a cascade of modifications across various aspects of the product.

Ripple Effects on Design and Development

When specifications change, they rarely do so in isolation. A modification in one component can affect the entire system, leading to a need for redesigning other components or even reevaluating the entire product architecture. For example, changing the battery in a device for higher capacity might necessitate alterations in the device's size, weight distribution, and heat dissipation system. These changes can be complex and time-consuming, requiring careful analysis and testing to ensure they don't compromise the product's functionality or integrity.

Impact on Project Timeline and Costs

Each change in specifications can significantly impact hardware project timeline. Redesigning, retesting, and ensuring that the new design still meets all regulatory and compliance requirements can add weeks or even months to the development schedule. Moreover, these changes often lead to increased costs – not just in the form of additional materials and labor, but also in potential delays to market entry, which can affect revenue projections and market competitiveness.

Balancing Agility and Stability

The key challenge for hardware developers is to balance the agility needed to incorporate valuable feedback and the stability required to maintain progress toward project milestones. This balancing act involves:

Effective Change Management: Implementing robust change management processes that evaluate the impact of specification changes before they are adopted.

Iterative Development: Adopting an iterative approach to design and development, which allows for more flexibility to accommodate changes without derailing the entire project.

Stakeholder Communication: Maintaining clear and frequent communication with all stakeholders, including team members, suppliers, and customers, to ensure that changes are well-coordinated and understood.

Strategic Decision-Making

Each decision to alter specifications must be strategic. It involves weighing the benefits of the change against the costs and delays it may incur. The decision-making process should consider not only the immediate implications of the change but also its long-term effects on product performance, user satisfaction, and market success.

Hence, specification evolution is a natural part of the hardware development process. While it brings opportunities for improvement and alignment with market needs, it also poses challenges in terms of project management and resource allocation. Successful navigation of these challenges requires a strategic approach to change management, an adaptable development methodology, and a comprehensive understanding of the interdependencies within the product design.

Financial Bottlenecks

Financial resources are not just fuel, they are the very lifeblood that sustains the project. The stability of funding is a critical factor, often determining the pace and momentum of the entire development process. Financial bottlenecks, characterized by a scarcity of funds, present a substantial challenge, capable of halting progress and, in severe cases, jeopardizing the project's viability.

Supply Chain Management: Dealing with Component Unavailability

Effective supply chain management is crucial in the context of hardware development. The availability of components is a cornerstone that can either accelerate progress or bring it to a grinding halt. Dealing with the unavailability of components, a challenge often encountered in this domain, requires strategic foresight and adept management to mitigate its impact on the project schedule.

The Criticality of Component Availability

Hardware development is deeply intertwined with the supply chain, relying on a myriad of components, from basic nuts and bolts to sophisticated electronic parts. The unavailability of even a single element can stall production, impacting everything from prototyping to final assembly. This disruption is not merely an inconvenience; it is a significant bottleneck that can derail the planned development timeline.

Consequences of Component Shortages

Component shortages can have a cascading effect on the hardware development process:

Delayed Prototyping and Testing: The absence of essential components can delay the prototyping phase, postponing crucial testing and iteration cycles.

Extended Development Timelines: Each delay in the supply chain extends the overall development timeline, pushing back milestones and potentially delaying market entry.

Increased Costs: Scarcity of components often leads to increased prices, impacting the overall budget of the project. In some cases, it might necessitate sourcing from alternative suppliers at higher costs.

Project Uncertainty: Frequent and unpredictable delays in component availability can create a sense of uncertainty, affecting stakeholder confidence and investment.

Strategies for Supply Chain Resilience

To overcome the challenges posed by component unavailability, hardware developers must employ strategic supply chain management:

Diversified Sourcing: Relying on a single supplier for components is risky. Developing relationships with multiple suppliers can provide alternatives when one source fails.

Inventory Management: Maintaining a buffer stock of critical components can help mitigate the impact of short-term shortages.

Flexible Design: Where possible, designing hardware that can accommodate components from multiple manufacturers can provide flexibility in sourcing.

Real-time Supply Chain Monitoring: Implementing systems to monitor the supply chain in real-time allows for early identification of potential shortages and quicker response times.

Strong Supplier Relationships: Building robust relationships with suppliers can lead to better communication, reliability, and prioritization in times of scarcity.

Contingency Planning: Having a well-defined contingency plan for component shortages is crucial. This might involve identifying alternative components or resequencing development activities to work on aspects of the project that are not affected by the shortage.

The Absence of a Contingency Plan in Hardware Development

In the complex and often unpredictable world of hardware development, the maxim 'hope for the best, but prepare for the worst' is particularly apt. The absence of a contingency plan, or 'Plan B', can be a critical oversight, leaving projects vulnerable to unforeseen challenges that could significantly derail progress. A robust contingency plan is not just a safety net; it's an integral part of strategic project management.

Understanding the Role of a Contingency Plan

A contingency plan in hardware development is a comprehensive strategy designed to address potential risks and unforeseen issues. It's a blueprint for action when the original plan encounters obstacles. This plan is essential because, in hardware development, numerous factors – from supply chain disruptions to technical challenges – can impact the project's success.

Developing a contingency plan involves several key elements:

Risk Assessment: Identifying potential risks and their impact on the project. This includes everything from technical failures to market changes.

Alternate Strategies: For each identified risk, developing alternative strategies and action plans. This might involve sourcing components from different suppliers, implementing backup manufacturing processes, or adjusting project timelines.

Resource Allocation: Pre-allocating resources, including funds and manpower, to manage potential risks effectively.

Communication Plan: Establishing clear communication channels to quickly disseminate information and instructions in case of an emergency.

Regular Review and Adaptation: Continuously monitoring the project's progress and adjusting the contingency plan as necessary to respond to new risks and challenges.

The Necessity of Flexibility and Adaptability

Flexibility and adaptability are at the heart of effective contingency planning. Hardware development projects are dynamic, with changing circumstances and new information. A contingency plan must be flexible enough to accommodate these changes, providing guidance that can be adapted to a range of scenarios.

From Delays to Launch: Crafting Your Personalized Action Plan

Going from initial delays to a successful launch is an art that combines strategic planning, adaptability, and meticulous execution. The creation of a personalized action plan is essential in this process, providing a roadmap that guides hardware developers through the complexities of bringing their vision to fruition.

The Essence of a Personalized Action Plan

A personalized action plan in hardware development is a detailed, step-by-step guide tailored to the unique challenges and objectives of the project. It transcends a generic checklist; it is a dynamic blueprint that aligns with the specific nuances of the product, team, and market environment.

Components of an Effective Action Plan

Clear Objectives and Milestones: The plan must outline clear, achievable objectives for each phase of the project, from design and prototyping to manufacturing and launch. Setting specific milestones helps in tracking progress and maintaining focus.

Risk Management Strategies: Incorporate comprehensive risk management strategies that identify potential obstacles and outline proactive measures to mitigate them. This includes both technical risks and external factors like market changes or supply chain disruptions.

Resource Allocation: Detail the allocation of resources, including budget, personnel, and time, ensuring that each aspect of the project is adequately resourced.

Flexible Timelines: While maintaining a schedule is crucial, the plan should allow for flexibility. Adaptive timelines that can accommodate unforeseen delays without derailing the entire project are essential.

Quality Control Measures: Implement rigorous quality control measures at each stage of development to ensure the product meets all required standards and specifications.

Stakeholder Engagement: Maintain a plan for regular communication and engagement with all stakeholders, including team members, suppliers, investors, and customers. This ensures alignment and support throughout the project lifecycle.

Contingency Plans: Integrate contingency plans that address potential delays and challenges. This includes alternative manufacturing options, backup suppliers, and strategies to address funding shortfalls.

Marketing and Launch Strategies: Develop a comprehensive marketing and launch strategy that aligns with the product's entry into the market. This includes pre-launch marketing activities, launch events, and post-launch support plans.

Adapting the Plan to Changing Circumstances

A successful action plan is not set in stone. It must be regularly reviewed and adapted to reflect the current status of the project, market trends, and any new insights or challenges that emerge. This iterative approach ensures that the plan remains relevant and effective throughout the development process.

As we close the chapter on overcoming delays and meticulously crafting action plans in hardware development, we stand on the threshold of an exciting new chapter. Here, we will delve deeply into the various stages of the product development lifecycle, uncovering the richness and complexity of each step.

Chapter 03

The Product Development Life Cycle

Imagine the journey of the Apollo Guidance Computer (AGC), an epitome of human ingenuity and a critical component of the Apollo missions that successfully landed humans on the moon. This marvel of hardware product development embarked on its journey in the early 1960s, at a time when the idea of space travel was as vast and unknown as space itself. It began as a mere concept, a visionary solution to the colossal challenge of navigating the unfathomable vastness of space. The AGC's development was not just a project; it was a monumental quest that encapsulated the very essence of the product development life cycle.

The Genesis: From Concept to Reality

The journey commenced with an audacious idea – to create a compact, yet powerful computer that could guide, navigate, and control the spacecraft during its lunar mission. It was an era when computers filled entire rooms, yet the AGC needed to be small, reliable, and extraordinarily capable. The ideation phase was marked by brainstorming sessions, theoretical calculations, and the convergence of various scientific disciplines. It was a time of boundless imagination and conceptualization.

Design and Innovation: Crafting the Brain of Apollo

As the concept took shape, the design phase of the Apollo Guidance Computer (AGC) saw teams of engineers and scientists at MIT and NASA meticulously crafting its architecture. This process was marked by a blend of groundbreaking and incremental innovations. While they ventured into uncharted territories, creating new integrated circuit technology and pioneering the development of silicon chips, they also embraced subtler, yet impactful, innovations.

One such example of incremental innovation, paralleling our work in fiber distribution, involved reimagining the design of fiber distribution hubs. Traditionally placed alongside roads, these hubs required access to both the front and back of the fiber panel. The existing design featured a hinge-type opening, which was functional but had limitations. Our innovation was a seemingly simple yet highly effective modification: converting the fiber panel from a hinge opening to a central axis rotation. This change significantly enhanced convenience and accessibility, allowing for easy maintenance in various weather conditions, including rain and snow. The panel remained protected within its enclosure, a crucial factor in its durability and reliability.

Testing and Prototyping: Ensuring Unprecedented Reliability

The AGC underwent a series of rigorous tests and refinements. Prototypes were built and subjected to conditions simulating those in space – extreme temperatures, vacuum, and intense radiation. Each test was a step towards perfection, ensuring that every component of the AGC could perform reliably in the most demanding circumstances. The relentless testing regime was a testament to the commitment to quality and precision.

Launch and Beyond: A Legacy of Innovation

Finally, the Apollo Guidance Computer made its journey to the moon, flawlessly guiding astronauts to lunar landings and safely back to Earth. It was a triumph of human intellect and technological prowess, a testament to the power of a well-executed product development life cycle.

The AGC's development journey, marked by groundbreaking innovation, meticulous design, and relentless testing, serves as a timeless example of what can be achieved when visionaries dare to dream and teams come together to turn those dreams into reality. As we delve further into the product development life cycle, let the story of the Apollo Guidance Computer inspire us to push the boundaries of what is possible.

Just as the Apollo Guidance Computer's journey epitomized the pinnacle of technological innovation and precision, the product development life cycle represents the journey every hardware innovation undergoes, from a nascent idea to a market-ready entity. It's a voyage not just of creation but of meticulous refinement and strategic execution.

Introducing the Product Development Life Cycle

Just as the Apollo Guidance Computer's development was marked by groundbreaking innovation and meticulous attention to detail, the product development life cycle is a testament to the power of strategic planning, creative thinking, and relentless pursuit of excellence. Each stage, significant in its own right, contributes to the ultimate goal of launching a successful, market-ready product. Each phase of this journey plays a critical role in shaping the final product, not just in its physical form but in its potential to meet and exceed market expectations.

Stages Overview: Charting the Course from Concept to Market-Ready Product

Some of the subjects covered in earlier chapter are again touched upon here, to emphasize on their importance and relevance.

The odyssey begins with the spark of concept development. Here, innovators and thinkers gather, bringing forth ideas fueled by imagination and necessity. This stage is about understanding the problem or identifying the market gap the product aims to address. It's where the foundation is laid, and the vision for the product starts taking shape.

As the journey progresses, the life cycle moves into design and prototyping. Here, the abstract concept begins its transformation into a tangible prototype. This phase is characterized by experimentation, iteration, and refinement. It's a stage of discovery, where challenges are met with creative solutions and where the product's form and function are defined and redefined.

Testing and validation follow, a phase where the product is put through rigorous trials to ensure it meets all necessary specifications, safety standards, and quality benchmarks. This stage is crucial in building trust and credibility, ensuring that the product is not only effective but also reliable and safe for consumer use.

Finally, the product enters the manufacturing and launch phases. This is where the product is produced at scale and introduced to the market. It's a stage of culmination, where all the hard work, innovation, and perseverance come to fruition.

Lifecycle Significance: Navigating the Path to Success

Understanding the significance of each stage in the product development life cycle is crucial. Each phase, from ideation and concept development to final launch, is integral in ensuring the product's success. They are stepping stones that must be carefully navigated to ensure that the final product not only meets the initial vision but also resonates with the market and consumers.

Innovation Integration: The Lifeline of the Product Development Journey

Throughout the product development lifecycle, innovation is not just a companion but a driving force, steering each stage with creativity and practical solutions. This journey of innovation is exemplified in our project with a global network service provider, where we were tasked to develop a Fiber Distribution Terminal for multi-dwelling installations, intended for Fiber To The Home applications.

In the early stages, the challenge was formidable: to create an innovative, easy-to-use, and foolproof fiber distribution terminal. The technical demands were stringent, with required insertion loss at 0.3 dB and return loss at 65 dB, while the existing benchmarks were only at 0.5 dB and 60 dB respectively. This phase was about dreaming big, thinking outside the box, and stretching the bounds of what was technically feasible.

As the product progressed, innovation took a more applied form. Undertaking rigorous R&D, we not only met but exceeded the demanding performance criteria, achieving the desired insertion loss and return loss. This phase of the lifecycle was about transforming lofty ideas into practical, real-world solutions that addressed specific market needs and technical challenges.

In the final stages, our innovative approach continued to influence manufacturing strategies and market introduction. This high-risk approach of breaking new ground in technical performance created a near-monopoly situation in the market, opening up significant business opportunities. It's a testament to how innovative strategies, from conception to manufacturing, can profoundly impact a product's acceptance and success in the market.

Concept Development: The Genesis of Innovation

In the realm of hardware product development, concept development is where the seed of innovation is planted and nurtured. It's a phase that demands not just creativity but also a strategic understanding of the market and the potential user base. This stage is about more than just ideation; it's about laying a solid foundation for the product that will eventually come to life.

Domain Identification: Carving Out Your Niche

The initial step in concept development is pinpointing your domain, a process that necessitates leveraging expertise to identify a niche where your product can flourish. This task demands a comprehensive understanding of both market demands and technological possibilities. Entrepreneurs are tasked with discerning where their strengths lie, and how these can be harnessed to fulfill unmet market needs or enhance existing solutions.

For example, the story of Fitbit serves as an illustrative case. Fitbit successfully identified a niche in the burgeoning field of wearable technology, specifically for fitness tracking. By tapping into the growing trend of health consciousness and integrating advances in sensor technology, Fitbit established itself as a trailblazer and leader in its domain.

Similarly, we embarked on an innovative venture in the realm of fiber optic network security, an area that was relatively uncharted for many telecom operators at the time. We identified the potential for intrusion in fiber networks, a significant but unacknowledged threat. Our approach combined creating an AI-integrated physical system to prevent such intrusions and educating potential clients about this latent threat through safe, controlled demonstrations of network vulnerabilities.

This venture was not just about introducing a new product; it was about pioneering a market for intrusion protection in fiber optics. We had to step out of our comfort zone, merging technological innovation with market education to create awareness and demand for our intrusion-proof system.

The success of this initiative was a collective effort, and it's crucial to acknowledge the significant contributions of the management and team members and the entire team. Their support was vital in navigating this new terrain, demonstrating the importance of not only identifying a niche but also committing to developing it, often by moving beyond conventional comfort zones.

Market Research: Ensuring Uniqueness and Need

Thorough market research is the cornerstone of successful concept development, it involves a deep dive into existing products, emerging trends, potential customer segments, and competitive landscapes. The goal is to ensure that the proposed concept is not just unique but also addresses a genuine need or desire in the market.

Effective market research might involve surveys, focus groups, and analysis of consumer behavior and market data. It's about gathering and interpreting information that can validate the need for your product and guide the development process. For example, when Dyson developed its first bagless vacuum cleaner, extensive market research helped them understand the frustrations consumers faced with traditional vacuums and guided them in designing a product that addressed those pain points.

Stakeholder Engagement: Refining the Concept through External Expertise

Engaging with stakeholders early in the concept development phase is crucial for refining and validating the product concept. Stakeholders can include potential customers, industry experts, potential partners, and even regulatory bodies. Their input can provide valuable insights into the feasibility, desirability, and viability of the product concept.

Engaging with stakeholders can take many forms, from formal consultations and partnerships to participatory design sessions. It's about opening up the development process to external expertise and perspectives, which can help in identifying potential issues early on and provide innovative ideas for improving the concept.

For example, the development of the Tesla electric car involved engaging with a wide range of stakeholders, from battery technology experts to potential consumers interested in sustainable technology. This engagement helped Tesla refine its concept into a product that stood out in the market for its innovation and quality.

Proof of Concept: Validating the Vision

The Proof of Concept (PoC) phase is a critical checkpoint in the hardware product development journey. It's where the theoretical meets the practical, and ideas are subjected to the first real-world tests. This stage is about demonstrating feasibility, not just in terms of technical capability but also in practicality and market potential.

Feasibility Demonstration: Testing the Waters

Demonstrating the feasibility of a concept involves a series of steps designed to validate the core idea. It's about answering fundamental questions: Can this product be made? Does it perform as expected? Is it solving the problem it's meant to solve? This phase might involve creating a basic prototype, conducting simulations, or implementing a scaled-down version of the product.

For instance, in the development of the first digital camera by Kodak, the feasibility demonstration was a crucial step. It involved building a working prototype that proved the concept of digital photography, even though the initial version was far from market-ready. This proof of concept was vital in showcasing the potential of digital imaging technology.

Resource Utilization: Smart Allocation for Maximum Insight

Effective resource utilization is about leveraging the tools, technologies, and expertise at your disposal to test the concept efficiently and effectively. This might involve using simulation software to model product behavior, repurposing existing technology to build a prototype, or utilizing shared spaces like maker labs that provide access to tools and equipment.

Efficient resource utilization is not just about saving money; it's about speed and learning. The quicker you can iterate and learn from PoC tests, the faster you can refine your concept. SpaceX's approach to rocket development exemplifies this, where rapid prototyping and testing have been key to their iterative design process, allowing them to quickly learn from failures and refine their designs.

Functionality Focus: Keeping an Eye on the Prize

During the PoC phase, it's essential to maintain a sharp focus on the product's intended functionality. This means prioritizing the core features that will deliver the most value to the user and ensuring these are the focus of your testing. It's about distinguishing 'must-have' features from 'nice-to-have' ones and ensuring the PoC validates these critical functionalities.

In the development of the original iPhone, the functionality focus was clear – to create a phone with a user-friendly touchscreen interface and an intuitive operating system. The PoC stages were heavily focused on ensuring these functionalities were not just feasible but performed to a high standard.

The Proof of Concept phase is where ideas begin to confront reality, and visions start to take tangible form. By carefully demonstrating feasibility, utilizing resources effectively, and maintaining a clear focus on functionality, entrepreneurs can navigate this phase effectively. It sets the stage for moving into more detailed design and development, armed with the confidence that the concept has been validated and is worth pursuing further.

Design Phase: Sculpting the Blueprint of Innovation

The design phase in product development is where the validated concept begins to take shape. It's a critical period of transformation, where the abstract idea morphs into a detailed plan ready for production. This phase requires a deep dive into the technicalities, aesthetics, and practicalities of bringing the product to life.

System Architecture: Crafting the Blueprint of the Product

Developing the system architecture is similar to drawing the master plan of a building. It involves outlining the product's structure, components, and the interrelationships between them. This is where decisions are made about the hardware's core functionalities, the software that will drive it, and how they will interact to create a cohesive, efficient system.

For instance, the development of the Nest Thermostat required meticulous system architecture planning. The team had to integrate various components like sensors, Wi-Fi modules, and user interface elements into a compact, efficient design that could learn and adapt to user behavior.

Design Considerations: Balancing Industrial Design with Technical Specifications

Balancing industrial design with technical specifications is crucial in creating a product that's not only functional but also appealing and user-friendly. Industrial design focuses on the aesthetics and user experience aspects, ensuring the product is ergonomic, visually pleasing, and intuitive to use. Technical specifications, on the other hand, deal with the product's functional aspects and performance criteria. As an example a metal strip on the product cover can deteriorate the WiFi range performance.

This stage often involves collaboration between designers and engineers, ensuring that the product's look and feel are in harmony with its technical capabilities. Apple's approach to product design exemplifies this balance. The company is known for its sleek, minimalist designs that seamlessly integrate with advanced technology, as seen in products like the MacBook.

Resource Management: Deciding between In-House Development and Outsourcing

Deciding where and how to develop the product is a critical part of the design phase. Resource management involves determining whether to develop in-house, outsource, or a combination of both. In-house development offers greater control over the process and can be beneficial for highly innovative or secretive projects. Outsourcing, however, can provide access to specialized skills and potentially lower costs.

The choice depends on various factors, including the company's expertise, budget, time constraints, and the complexity of the product. For example, many tech companies, including Google with its Pixel smartphones, opt for a combination of in-house and outsourced development to leverage external expertise while maintaining control over critical design and innovation aspects. It is to be carefully noted that a strong Intellectual Property right agreement is in place while outsourcing the designs.

The design phase is a pivotal stage in the hardware product development cycle, transforming the proven concept into a detailed blueprint ready for prototyping and production. By meticulously crafting the system architecture, balancing industrial design with technical specifications, and strategically managing resources, companies can navigate this phase effectively, setting the stage for a successful product launch. This stage is about precision, vision, and strategic decision-making, ensuring that the final product will meet the market's needs and exceed user expectations.

Alpha and Beta Prototypes: The Path from Concept to Reality

The evolution from design to prototype marks a pivotal transition from the theoretical to the tangible. This stage is where concepts and plans are transformed into physical models that can be interacted with, tested, and refined. The alpha and beta prototype phases are critical for testing the product's real-world functionality and making necessary adjustments before it goes into mass production.

Prototype Development: Crafting the Initial and Refined Models

Prototype development initiates with the creation of alpha prototypes. These initial models are often rudimentary, designed to test and validate the core functionality of the product. They provide the first opportunity to assess the design in a physical form and identify any technical or practical issues. The alpha prototypes are less about aesthetics and more about understanding the product's workings at a fundamental level.

Once the alpha prototype has been evaluated and refined based on initial testing, the process advances to the beta prototypes. These versions are more sophisticated, incorporating all the design elements and functionalities expected in the final product. Beta prototypes are typically used for more extensive testing, often in real-world settings with potential users. This stage is crucial for gathering detailed feedback on the product's performance, usability, and appeal to the target market.

In the development of complex hardware, such as medical devices or aerospace components, the journey from alpha to beta prototypes is marked by rigorous testing and validation. For example, in aerospace, each component of a prototype might undergo stress tests, thermal simulations, and performance assessments to ensure it can withstand the extreme conditions of flight. The iterative process of refining these prototypes is guided by precise industry standards and regulatory requirements, ensuring the final product is not only innovative but also safe and reliable.

User Feedback Integration: Fine-Tuning the Product

Incorporating user feedback is an indispensable aspect of the prototype phase. Feedback from real users provides insights into how the product is used, identifies any issues or challenges encountered, and suggests possible improvements. Engaging with users during the beta phase helps ensure that the final product will meet their needs and expectations, enhancing its marketability and user satisfaction.

For instance, in the development of ergonomic tools or equipment, user feedback can highlight issues with the design that might not be apparent in a lab setting. This might include the tool's weight, balance, or ease of use in various conditions. By integrating this feedback, designers can make targeted improvements that significantly enhance the product's usability and appeal. An example is a handheld scanner device which is used in cold storages. The components within it have to withstand the extreme low temperatures, there are possibilities of higher electro static shocks, etc. Thus, the device has to be well designed considering these working conditions.

Iterative Refinement: Perfecting the Product through Successive Models

Iterative refinement is the essence of the prototype development phase. It involves a continuous cycle of testing, feedback, improvement, and retesting. Each iteration refines the product, improving its design, functionality, and user experience. This process is fundamental to transforming a good concept into an exceptional product. It requires openness to learning, willingness to adapt, and a commitment to pursuing excellence.

Through iterative refinement, developers can gradually resolve any issues, enhance the product's performance, and ensure that every aspect is optimized for the end user. This attention to detail and dedication to continuous improvement is what differentiates a successful product in the competitive hardware market.

Pilot Build and Regulatory Compliances: Final Preparations Before Market Entry

As products transition from refined prototypes to market-ready entities, they enter a critical phase known as the pilot build. This phase is coupled closely with the necessity of understanding and adhering to regulatory compliances—a stage that ensures not only the product's functionality and quality but also its safety and legality.

Pilot Production: The Crucial Test Run

Pilot production is essentially a test run of the manufacturing process on a smaller scale. This stage allows manufacturers to identify any potential issues in the production line, from material sourcing and machinery operation to assembly and packaging. It's about ensuring that when full-scale production commences, every aspect of the manufacturing process is optimized for efficiency, quality, and consistency.

During pilot production, small batches of the product are created to test and validate the entire production process. It's a stage where theoretical production plans meet the practicalities of real-world manufacturing. Any issues encountered here can be addressed and resolved without the pressures and costs associated with full-scale production.

In India, where manufacturing sectors are burgeoning, conducting a successful pilot build means paying close attention to the supply chain and logistics, given the country's diverse and sometimes challenging geography. For example, when automotive companies like Maruti Suzuki introduce a new model, they undertake extensive pilot production to ensure that their extensive supplier network can consistently deliver parts that meet their strict quality standards.

Volume Manufacturing Prep: Scaling Up for the Masses

Transitioning from prototypes to mass production, volume manufacturing preparation is about fine-tuning the manufacturing process for efficiency, cost-effectiveness, and scalability. This stage involves optimizing the design for manufacturability, ensuring supply chain readiness, and setting up or refining assembly lines for larger production volumes. It's about ensuring that each unit produced maintains the same high quality as the prototypes and that production can ramp up to meet market demand.

An instance of this is the production ramp-up of the Tata Nano, once dubbed the world's cheapest car. To meet the anticipated high demand and keep costs low, Tata Motors had to rethink traditional car manufacturing processes. This involved setting up a new production plant and innovating in areas like material sourcing and assembly to achieve the desired scale and cost targets.

Launch Readiness: Dotting the I's and Crossing the T's

Launch readiness is the final checkpoint before the product hits the market. It's a comprehensive review to ensure that every aspect of the product and the manufacturing process is in line with what was planned and tested. This includes finalizing packaging, confirming regulatory compliances, ensuring quality control on the initial production run, and preparing for logistics and distribution.

Launch readiness also involves ensuring that the marketing, sales, and customer support teams are prepared to handle the product's introduction to the market. They must be equipped with the necessary information and resources to effectively promote the product, assist customers, and handle any post-launch issues.

The reliability evaluation and volume manufacturing phase are where a product's durability is ensured, and its transition to mass production is finalized. Rigorous reliability testing confirms the product's long-term performance, while careful preparation for volume manufacturing ensures it can be produced at scale efficiently and cost-effectively. As the product moves toward launch readiness, every detail is checked and rechecked, ensuring that when it finally reaches the market, it's fully prepared to meet the expectations and needs of consumers. This phase marks the culmination of the product development cycle, representing the transition from an innovative concept to a market-ready solution.

As we conclude our exploration of the foundational aspects of the product development life cycle, we stand on the threshold of a deeper journey into each individual stage. From the initial spark of Ideation to the rigorous processes of Proof of Concept, and through the meticulous steps of Design, Prototype Development, and beyond, the upcoming chapters will delve into the intricacies of each phase in great detail.

We'll explore the art and science behind transforming a nascent idea into a tangible, market-ready product. Each chapter will be dedicated to one stage of the product development cycle, offering in-depth insights, real-world examples, and strategic considerations. As we move forward, let's carry with us the lessons learned from the overview. The journey ahead is one of discovery, challenge, and innovation.

Ideation – The Birthplace of Innovation

Let's begin the journey of the ideation process by recalling the groundbreaking story of the Wright brothers, Orville and Wilbur, who revolutionized transportation through the invention of the first successful airplane. Their relentless pursuit of flight began not with gears and engines, but with an idea, a vision that humans could fly. This iconic piece of hardware innovation was born out of an ideation process that combined curiosity, observation, and a series of methodical, creative steps. Their journey exemplifies the power and importance of ideation in hardware product development, where a single idea can lift humanity into a new era.

Stages of Ideation: Structuring the Creative Process

The ideation process, often perceived as a spontaneous burst of creativity, is actually a structured journey that can be broken down into manageable parts. Understanding these stages is crucial for innovators looking to navigate the complex path from conception to realization.

Understanding the Problem: The first stage involves immersing oneself in the problem space. For the Wright brothers, it was the challenge of controlled, powered flight. They spent countless hours studying birds in flight, understanding the mechanics of aerodynamics, and identifying the core problem—how to control an aircraft in the air.

Divergent Thinking: Once the problem is understood, the next stage is divergent thinking. This involves exploring as many solutions as possible, often in a brainstorming session where quantity trumps quality. It's about letting the imagination run wild without the constraints of practicality or feasibility. The Wright brothers, for instance, considered numerous possibilities, from different wing shapes to various methods of controlling the craft.

Convergent Thinking: After a plethora of ideas have been generated, the process shifts to convergent thinking. This stage is about evaluating, critiquing, and selecting the most promising ideas for further exploration. It's a funneling process where the vast field of ideas is narrowed down to a few viable options. The Wright brothers, through their meticulous research and understanding of aerodynamics, converged on the idea of a three-axis control system, which became a fundamental principle of flight.

Prototyping and Feedback: The selected ideas are then turned into prototypes. This stage is about bringing the abstract into the tangible world, creating a model or a simulation of the product. It's also about testing these prototypes and gathering feedback, which is crucial for refining the idea. The Wright brothers built numerous gliders, testing and refining their designs based on the performance of each prototype.

Finalization: The final stage of ideation is the selection and finalization of the concept that will be developed into a full-fledged product. It's about taking the prototype that has shown the most promise and preparing it for the next stages of development. For the Wright brothers, this meant choosing the best design for their powered aircraft, which would eventually lead to the historic first flight.

The ideation process, as illustrated by the Wright brothers' journey, is a structured yet creative endeavor that turns abstract problems into groundbreaking hardware innovations. By understanding and navigating the stages of ideation—from understanding the problem to finalizing the concept—innovators can set the stage for successful product development.

Techniques and Tools: Harnessing Creativity to Capture Innovation

Building on the foundational stages of ideation, as exemplified by the Wright brothers' journey to flight, we transition to the tools and techniques that fuel this creative process. Just as the brothers utilized sketches, models, and wind tunnels to shape their airborne aspirations, modern innovators have a plethora of methodologies at their disposal to generate and capture ideas. These techniques are the catalysts that transform a spark of possibility into a blaze of innovation.

Divergent Thinking Tools: Expanding the Creative Horizon

Brainstorming Sessions: Perhaps the most recognized tool for idea generation, brainstorming is about gathering a group of people and collectively exploring a problem. The key is to encourage free-flowing thoughts and defer judgment, allowing even the most outlandish ideas to surface. This technique was fundamental to the Wright brothers as they exchanged and built upon each other's thoughts.

Mind Mapping: This visual tool involves creating a diagram that represents ideas, words, tasks, or other items linked to and arranged around a central concept. It's particularly effective for visual thinkers, helping to organize and structure thoughts. For hardware developers, mind mapping can be a powerful tool to visually breakdown complex systems and explore connections between various components.

SCAMPER: An acronym for Substitute, Combine, Adapt, Modify, Put to another use, Eliminate, and Reverse, SCAMPER is a checklist-based technique that encourages thinkers to ask questions about existing products or ideas to spark new ones. It's about looking at things differently and exploring various perspectives.

Convergent Thinking Tools: Refining and Selecting Ideas

SWOT Analysis: Standing for Strengths, Weaknesses, Opportunities, and Threats, SWOT Analysis is a strategic planning tool used to evaluate an idea. It helps innovators understand the internal and external factors that could impact the success of their idea, guiding them to refine and improve it.

Pugh Matrix: A decision-matrix method used to determine the best solution out of a set of options based on multiple criteria. In hardware development, it can be instrumental in comparing different design iterations and selecting the one that meets the most critical criteria.

Six Thinking Hats: Developed by Edward de Bono, this tool encourages thinkers to look at a problem from six different perspectives (emotional, factual, creative, positive, negative, and organizational). It's a method that ensures all aspects of a problem are considered, leading to more thorough and holistic decision-making.

Idea Capture Tools: Ensuring No Thought is Lost

Idea Management Software: Platforms like Trello, Asana, or Jira allow teams to capture, track, and prioritize ideas. These tools are particularly effective in collaborative environments where ideas are constantly being generated and need to be organized.

Digital Note-Taking: Apps like Evernote or OneNote allow innovators to capture thoughts as they occur, ensuring that no burst of inspiration is lost. They offer the flexibility to record ideas in various formats, be it text, voice, or image.

Physical Idea Journals: Sometimes, the simplicity of a notebook is all that's needed. Many inventors and designers prefer sketching or writing in physical journals, finding that the act of writing by hand stimulates creativity.

Just as the Wright brothers iteratively refined their designs, the modern innovator must use a blend of divergent and convergent thinking tools to generate and refine their ideas. Coupled with effective idea capture tools, these methodologies ensure that no potential innovation slips through the cracks.

Cross-Industry Inspiration: Diversifying the Sources of Innovation

Cross-industry inspiration entails looking beyond the familiar confines of your sector to draw innovative ideas from different fields. This approach infuses fresh perspectives into your ideation process and sparks unconventional solutions that can revolutionize your products.

Drawing from a Diverse Innovation Palette

Learning from Other Industries: Every industry has its unique challenges and innovative solutions. By studying how different sectors tackle their problems, you can uncover insights and approaches that can be adapted to your context. For instance, the healthcare sector's use of telemedicine technology to reach remote patients can inspire ways to provide remote diagnostics for machinery in the manufacturing industry. In my earlier profession, where an innovative concept of a fiber distribution hub for multi dwelling building was warranted, the idea got inspired from electrical extension boxes, fire hose handling method, etc. This was translated to US Patent.

Adopting Best Practices: Sometimes, the most innovative ideas come from applying best practices from one field to another.

The concept of lean manufacturing, originated in the automotive industry at Toyota, has now been adopted worldwide across various sectors to improve efficiency and reduce waste.

Technology Transfer: Innovations in technology often have applications that span multiple industries. For example, the GPS technology initially developed for military use has found its way into navigation systems, smartphones, and even farming equipment for precision agriculture. By keeping an eye on technological advancements in other fields, you can identify potential applications and integrations for your products.

Bridging Sectors for Hybrid Innovations

Collaborative Ventures: Collaborating with companies or research institutions from different sectors can lead to hybrid innovations that blend the best of both worlds. These partnerships can open up new markets and create products that are more versatile and innovative. For instance, partnerships between biotech firms and tech companies are driving innovations in wearable health monitors that track everything from heart rate to blood glucose levels.

Cross-Disciplinary Teams: Building teams with diverse professional backgrounds can naturally lead to cross-industry inspiration. When engineers, designers, scientists, and marketers collaborate, they bring different perspectives and problem-solving approaches, leading to more creative and well-rounded solutions.

Industry Conferences and Expos: Participating in or attending events that bring together professionals from various industries can be a goldmine for inspiration. These gatherings are opportunities to learn about the latest trends, technologies, and challenges faced by different sectors, sparking ideas for how these could be relevant to your own work.

Drawing inspiration from different sectors isn't just about borrowing ideas; it's about reimagining and integrating these concepts in a way that suits your product's unique needs and challenges. By stepping out of the industry silo and exploring the wide array of innovations across sectors, you can uncover surprising synergies and opportunities for your ideation process.

The Role of Research in Ideation: Unveiling Market Needs and Consumer Desires

Let's now focus on a fundamental aspect of ideation: research. Understanding the market's needs and consumer desires is crucial for ensuring that your ideas are not just creative but also relevant and viable. This phase of ideation is where intuition meets data, and creativity aligns with practicality.

Market Research: Decoding the Landscape of Opportunities

Trend Analysis: Staying ahead of the curve requires an understanding of current and emerging market trends. This involves analyzing industry reports, market forecasts, and trend predictions. By identifying what's on the horizon, you can position your product to meet future demands. Consider how Dyson, the technology company best known for its vacuum cleaners, continually innovates by analyzing and predicting market trends. Dyson's move into air purifiers and hand dryers wasn't random but a result of understanding broader trends about consumer health consciousness and environmental sustainability.

Competitor Analysis: Understanding what your competitors are doing can provide valuable insights into market gaps and opportunities. It's not just about what they're doing right but also where they're lacking. This analysis can inspire innovations that fill those gaps or offer better solutions. Dyson's rise was also partly due to its meticulous study of existing products in the market. By understanding and dissecting the flaws in traditional vacuum cleaners, Dyson was able to innovate and introduce the bagless vacuum, fundamentally changing the industry.

Segmentation and Targeting: Not all markets are monolithic. Identifying different market segments and understanding their specific needs and preferences can help tailor your ideation process. This might involve creating personas or detailed profiles of potential users, helping to focus your ideas on solving real and specific problems. Dyson's approach to identifying specific market segments, such as pet owners or allergy sufferers, and then designing products to meet these groups' specific needs is a testament to the power of targeted market research.

Consumer Research: Delving into the Heart of User Needs

Surveys and Interviews: Direct feedback from potential users is invaluable. Surveys and interviews can reveal what consumers truly want, their pain points, and their unmet needs. This direct line of communication ensures that your ideation process is grounded in real-world desires and problems.

Ethnographic Studies: Sometimes, to truly understand consumers, you need to observe them in their natural environment. Ethnographic studies involve observing how potential users interact with products or services in their daily lives, providing deep insights into their behaviors, needs, and frustrations.

Feedback Loops: Establishing early and continuous feedback loops can keep your ideation process aligned with consumer needs. This might involve creating prototypes or mock-ups and getting user feedback or using platforms to crowdsource ideas and gauge consumer interest.

Leveraging Research for Informed Ideation

Data-Driven Decisions: Armed with research, you can make informed decisions about which ideas to pursue. Data provides a reality check for your creativity, ensuring that your innovations are not just feasible but also desired by the market. Tesla's approach to car manufacturing is heavily data-driven. By continually analyzing customer feedback, market trends, and performance data, Tesla iteratively improves its vehicles' designs and functionalities.

Identifying Unmet Needs: Research can reveal gaps in the market—needs that are currently unmet or poorly served. These gaps represent golden opportunities for innovation, where new or improved products can make a significant impact. The development of Tesla's Autopilot system was in response to an identified need for safer and more comfortable driving experiences. Extensive research into driving habits and existing autonomous technologies allowed Tesla to innovate in a way that significantly set them apart from competitors.

Customizing Solutions: Understanding consumer demographics, behaviors, and preferences allows for customization of solutions. This might involve tailoring products for different age groups, lifestyles, or cultural contexts, ensuring that your product resonates with its intended audience.

Technological Trends: Harnessing the Future of Innovation

As we move from the broad scope of market and consumer research to the more focused realm of technological trends, it's vital to understand how staying ahead of tech curves and integrating emerging technologies can significantly enhance the ideation process. In a world where technology evolves at breakneck speed, being attuned to the latest advancements is essential for maintaining a competitive edge and fostering innovation.

Navigating the Ever-Changing Technological Landscape

Continuous Learning: The first step in staying ahead of technological trends is fostering a culture of continuous learning and curiosity. This means regularly engaging with tech news, attending industry conferences, and participating in forums and discussions. It's about keeping your finger on the pulse of what's new and what's next.

Technology Scouting: Companies often engage in technology scouting to identify emerging technologies that could impact their industry or be harnessed for new product development. This proactive approach involves analyzing patents, academic research, and market reports to predict which technologies will be game-changers.

Collaboration with Tech Innovators: Building relationships with universities, research institutions, and startups can provide early insights into cutting-edge technologies. These collaborations can lead to partnerships that help integrate these advancements into your products.

Integrating Emerging Technologies: The Apple iPhone's Revolutionary Approach

A prime example of harnessing technological trends is Apple's development of the iPhone. When the first iPhone was introduced, it wasn't just a new product; it was a culmination of several emerging technologies seamlessly integrated into a single device. Apple didn't invent the touchscreen, the smartphone, or even the internet-connected device, but they foresaw the potential of these technologies converging into a user-friendly, multi-functional device.

The iPhone's development involved:

Touchscreen Technology: Apple recognized the potential of capacitive touchscreens to provide a more intuitive and direct interface for users.

Miniaturization: Advancements in miniaturization allowed Apple to pack more functionality into a smaller space, making the iPhone a powerful computer that fits in your pocket.

Wireless Connectivity: By leveraging emerging wireless technologies, Apple ensured that the iPhone could connect anywhere, anytime, changing how people interact with the internet and each other.

This integration wasn't just about using the latest tech, it was about understanding how these technologies could work together to create a revolutionary product. The iPhone's success was due in large part to Apple's ability to anticipate and incorporate technological trends, setting a new standard for what a mobile device could be.

Validating Your Idea

Transitioning from the exploration of technological trends and how they can shape innovative products, we now turn our focus towards the crucial step of validating your idea. It's a phase where the theoretical meets the practical, and dreams are put through the rigors of reality. Validating your idea through feasibility studies is not just a checkpoint but a crucial process that can save time, resources, and guide your ideation towards success.

Feasibility Studies: The Crucial Litmus Test for Your Ideas

Technical Feasibility: Before an idea can become a tangible product, it's essential to assess whether it's technically possible. This involves a detailed analysis of the technology needed, the skills required to develop it, and the resources necessary to bring it to life. For instance, if you're considering developing a new type of battery, you'd need to research the chemistry involved, the manufacturing process, and whether you can access the materials needed.

Economic Feasibility: An idea might be technically possible but not economically viable. Economic feasibility studies look at the cost of developing and producing the product and compare it with the potential revenue. This analysis helps determine whether the idea can turn a profit. Consider the Concorde supersonic plane: technically feasible and a marvel of engineering, but economically, it struggled to be profitable due to the high costs of operation and limited passenger capacity.

Market Feasibility: Understanding the market demand for your product is crucial. This involves analyzing market size, potential customers, competition, and market trends. Market feasibility helps ensure that there's a desire and need for your product and that it has a place in the competitive landscape. For example, before launching a new smartwatch, a company would study the market to understand consumer preferences, what features are in demand, and how competitors are positioning their products.

Operational Feasibility: Even if an idea is technically and economically viable and there's a market for it, you need to consider whether you can operationally bring it to life. This involves assessing your company's capabilities, the potential need for partnerships, supply chain logistics, and the infrastructure required. It's about understanding the practicalities of turning the idea into a product that can be produced and delivered to consumers.

Legal and Ethical Feasibility: Your idea must adhere to legal standards and ethical considerations. This involves ensuring that the product complies with relevant regulations, obtaining the necessary licenses, and considering the ethical implications of your product on society and the environment. For instance, if you're developing a new health device, it would need to comply with medical regulations and standards, and you'd need to consider the privacy and ethical implications of collecting and using health data.

Feasibility studies are a multifaceted evaluation that looks at every aspect of turning an idea into a successful product. They are not a hurdle but a guiding tool that helps refine your idea, ensuring it's grounded in reality and has the potential for success.

Proof of Concept: Bridging the Gap Between Idea and Reality

As we transition from the essential groundwork of feasibility studies, we move into the tangible realm of transforming your validated idea into a concrete demonstration: the Proof of Concept (PoC). This stage is where the rubber meets the road, where theoretical viability is tested in the real world, and where ideas begin their transformation into tangible products. The PoC serves as a bridge, turning abstract concepts into physical models or prototypes that demonstrate the idea's potential and functionality.

The Journey from Conceptualization to Tangibility

Defining Objectives for Your PoC: The first step in developing a PoC is to clearly define what you aim to demonstrate or test. This involves setting specific, measurable objectives that align with the challenges and questions raised during the feasibility study. Whether it's proving that a certain mechanism works, testing a new material, or demonstrating a user interaction, your PoC should be designed with these objectives in mind.

Selecting the Right Approach: Depending on what you need to demonstrate, your PoC might take various forms, from simple models or simulations to more complex prototypes. For a hardware product, this could involve creating a scaled-down model, a fully functional prototype, or a digital simulation. The approach should provide the quickest and most cost-effective way to test your idea's core aspects.

Gathering Resources and Skills: Developing a PoC often requires a specific set of skills and resources. This might involve bringing in experts with specialized knowledge, accessing certain tools or technology, or collaborating with research institutions. For instance, creating a PoC for a new medical device might require collaboration with healthcare professionals to ensure it meets real-world needs and standards.

Testing and Iterating: Once your PoC is developed, it's time to test it against the objectives you've set. This stage is crucial for gathering insights about how your idea performs in a practical context. It's also a time for iteration; based on the feedback and results, you'll likely need to make adjustments and refinements. This iterative process is where your idea evolves and improves, moving closer to a viable product.

Documenting and Learning: Every PoC is a learning opportunity. Whether it succeeds in demonstrating what you intended or reveals new challenges, documenting every aspect of the development and testing process is crucial. This documentation will be invaluable as you move forward, helping to guide future development and avoid potential pitfalls.

An Example of PoC in Action: Consider the development of the first digital camera by Eastman Kodak in 1975. The initial prototype was bulky and had low resolution, but it demonstrated the fundamental concept that images could be captured digitally. This PoC was a pivotal moment, not just for Kodak but for the entire photography industry, setting the stage for the digital revolution.

Case Studies: Learning from the Trenches of Innovation

Let's now delve into narratives that provide invaluable insights into the ideation process. Through detailed case studies and an examination of common pitfalls, we can glean wisdom from those who've navigated the path of innovation before us.

Startup Spotlight: The Detailed Journey of Fitbit's Ideation

Fitbit's story is not just a tale of success but a blueprint of the ideation process executed effectively. It's a narrative that illustrates the journey from spotting a market need to overcoming the myriad challenges faced by startups. Let's delve deeper into each stage of Fitbit's ideation journey to understand how it transformed from a concept into a product that redefined personal fitness.

Phase 1: Spotting the Opportunity

Identifying the Gap: James Park and Eric Friedman recognized a burgeoning trend in health and fitness. People were becoming increasingly health-conscious but lacked a convenient way to track their daily activities. The existing products were either too complex or insufficiently integrated into users' lifestyles. This recognition of a market gap was the first spark of Fitbit's ideation process.

Envisioning a Solution: Park and Friedman envisioned a device that was not just a tracker but a companion in the user's fitness journey. They imagined a wearable device that was discreet, user-friendly, and provided real-time feedback. This vision was crucial; it wasn't about creating just another product but about fulfilling a specific consumer need in a way that hadn't been done before.

Phase 2: Shaping the Idea

Rapid Prototyping: Fitbit's ideation process was characterized by rapid prototyping. They developed multiple versions of the product, each time refining it based on feedback and their evolving understanding of the technology and market. This wasn't a linear process but an iterative one, with each prototype getting them closer to the ideal product.

Engaging Early Adopters: Early versions of Fitbit were shared with a select group of potential users. This early engagement was invaluable. It provided insights not just into what users wanted, but how they interacted with the device in real-life scenarios. This feedback was crucial in refining the product's design and functionality.

Phase 3: Overcoming Barriers

Technical Challenges: Developing a new type of wearable technology presented numerous technical challenges. From ensuring the device's accuracy to extending its battery life, each hurdle required innovative solutions. The team's technical prowess and willingness to experiment were crucial in overcoming these challenges.

Funding and Market Entry: Securing funding for a novel concept is always challenging. The founders pitched to numerous investors, faced rejections, and continued refining their pitch and product. Their persistence paid off when they received their first major funding, allowing them to move towards large-scale production.

Market Education: Introducing a new product category meant educating the market—not just about the product but about the concept of wearable fitness technology. Fitbit's strategy included clear communication about the benefits of the product and how it fit into the user's lifestyle, creating a new market for wearable fitness trackers.

Pitfalls to Avoid: Navigating Common Missteps in Hardware Startups

In the journey of transforming a groundbreaking idea into a tangible product, hardware startups often encounter a landscape riddled with potential pitfalls. Recognizing and understanding these common challenges is crucial for navigating the path successfully. Let's delve into these pitfalls, drawing lessons from those who've stumbled and providing strategies to avoid making the same mistakes.

Ignoring User Feedback: The Peril of a Closed Mind

Lesson Learned: Many startups, enamored by their vision, fall into the trap of ignoring or dismissing user feedback. This mistake was notably seen in the case of a startup that developed an innovative home gadget. Despite early user tests indicating that the device was too complex, the team pressed ahead, believing their vision was superior. Upon launch, the product's usability issues led to poor sales and negative reviews.

Strategy for Avoidance: Embrace a user-centric approach from the outset. Regularly collect and analyze feedback through surveys, focus groups, and prototype testing. Be open to criticism and willing to pivot or make changes based on what potential users are telling you. Remember, the market ultimately decides a product's success, not just the passion of its creators.

Underestimating Resource Needs: A Recipe for Shortfall

Lesson Learned: A common narrative among failed hardware startups is the underestimation of resources required for development, production, and market entry. One such startup in the robotics field realized too late that their funding was insufficient to cover the advanced materials and specialized labor needed, leading to a half-finished product and eventual shutdown.

Strategy for Avoidance: Conduct a thorough resource analysis early in the ideation phase. Consider all costs, including materials, labor, R&D, certification, and marketing. Seek advice from industry experts and mentors who can provide a reality check on your estimates. Plan for contingencies and ensure you have a buffer for unexpected costs.

Overcomplicating the Solution: Complexity as the Enemy of Innovation

Lesson Learned: In an effort to stand out, startups sometimes over-engineer their products, adding unnecessary features that complicate the user experience. A notable example was a startup that developed a kitchen appliance with a multitude of features, most of which were rarely used. The complexity made the product intimidating and ultimately less appealing to consumers.

Strategy for Avoidance: Adhere to the principle of simplicity. Start with a minimum viable product (MVP) that addresses the core problem effectively. Resist the temptation to add features until the basic functionality is proven and demanded by users. Complexity can always be introduced later, based on real-world usage and feedback.

Misreading the Market Map

Lesson Learned: Many startups have ventured forth with products they assumed the market needed, only to discover that there was no real demand or that they had misjudged the competitive landscape. A classic case involved a startup that developed a high-tech home security system, only to find that a similar, more affordable product was already available.

Strategy for Avoidance: Before diving deep into product development, invest time and resources in comprehensive market research. Understand your target audience, analyze your competitors, and identify your unique value proposition. Make sure there's a clear and compelling need for your product and that you have a strategy to differentiate it from existing solutions.

Ideation in Action

Transitioning from understanding the common pitfalls in hardware startups, let's focus on how to practically bring ideation into action. At the heart of this process are workshops and brainstorming sessions, dynamic platforms where ideas are born, shared, and refined. These sessions are not just meetings; they are orchestrated efforts to harness collective creativity and drive innovation.

Workshops and Brainstorms: The Engines of Ideation

Setting the Stage for Creativity: The first step in running an effective ideation session is creating the right environment. This means choosing a space that is comfortable and free from distractions. It might be a quiet room with whiteboards and sticky notes or an outdoor retreat. The key is to make participants feel relaxed and open to sharing their thoughts.

Diverse Participation: One of the strengths of workshops and brainstorming sessions is the diversity of ideas they can generate. Invite participants from different departments, backgrounds, and even outside your organization. For example, a hardware startup might include engineers, designers, marketers, and end-users in their sessions. Each group brings a unique perspective that can lead to more comprehensive and innovative solutions.

Structured Yet Flexible Approach: While creativity should flow freely, some structure is necessary to keep the session productive. This might involve setting a clear agenda, defining the problem or topic to be addressed, and allocating time for each activity. However, be prepared to adapt and follow the conversation where it leads. Sometimes, the most unexpected discussions can yield the most valuable insights.

Running Effective Ideation Sessions: Techniques and Tips

Brainstorming Techniques: Employ various brainstorming techniques to spur creativity. Methods like 'mind mapping' encourage participants to think beyond the obvious, while 'sketching sessions' can help visualize ideas. 'Reverse brainstorming,' where you think about how to cause the problem, can also lead to innovative ways to solve it.

Encouraging Participation: Ensure everyone has a chance to contribute. Techniques like 'round robin' where each person shares an idea in turn, can ensure quieter members have a voice. Another strategy is using anonymous idea submission, which can encourage more candid contributions.

Building on Ideas: Encourage participants to build on each other's ideas. Phrases like "Yes, and..." foster an atmosphere of collaboration and expansion, turning a single thought into a well-developed concept.

Capturing Everything: Document every idea, no matter how outlandish it may seem. Visual aids like whiteboards or digital tools can capture the session's output. These records are invaluable for future reference and can be the seeds for future innovations.

Ideo's Deep Dive Sessions

Ideo, a global design company known for its innovative approach, often conducts 'Deep Dive' sessions – intensive brainstorming sessions that bring together people from various disciplines to solve design challenges. These sessions have led to groundbreaking products like the first computer mouse for Apple and the stand-up toothpaste tube. The key to their success is a well-structured yet flexible approach, encouraging wild ideas, and a diverse group of participants.

User-Centered Design Thinking: Integrating User Insights into Ideation

User-Centered Design Thinking (UCDT) is a framework that keeps the user's needs, preferences, and behaviors at the forefront of the product development process. It's about empathizing with the users, understanding their challenges, and incorporating their feedback directly into the ideation phase to create solutions that truly resonate with them.

Principles of User-Centered Design Thinking

Empathy: At the heart of UCDT is empathy, the ability to understand and share the feelings of others. In the context of ideation, this means deeply understanding the user's world—their problems, their environment, and their desires. Techniques like user interviews, surveys, and observation are crucial for gathering these insights.

Defining User Needs: Once you've gathered user insights, the next step is to define the users' needs clearly and concisely. This often involves creating user personas and scenarios that depict the typical users and their interactions with the product. This definition provides a focused direction for ideation, ensuring that the ideas generated are relevant and targeted.

Ideation with the User in Mind: With a clear understanding of the user's needs, ideation sessions can be more directed and fruitful. Techniques like 'How Might We' questions help to frame the ideation challenges in a way that focuses on solving real user problems.

Incorporating User Feedback into Ideation

Continuous Feedback Loops: Establishing mechanisms for ongoing user feedback is crucial. This can be through prototypes, beta versions, or mock-ups shared with a user group for feedback. The key is to create a loop where user feedback is continually sought and used to refine the ideas and concepts being developed.

Co-Creation Sessions: Inviting users to participate in ideation sessions can provide direct and immediate insights. Users often can articulate their needs and challenges in ways that internal teams might not have considered. Co-creation can take the form of workshops, online forums, or focus groups where users and developers ideate together.

Rapid Prototyping and Testing: Creating quick and rough versions of products or features allows for immediate testing and feedback. This approach, part of the broader Lean UX methodology, ensures that the product is continually being shaped by user needs and feedback throughout the ideation phase.

The Development of OXO Good Grips

A classic example of incorporating user feedback into the ideation phase is the development of OXO Good Grips kitchen tools. The founder, Sam Farber, noticed that his wife, who had mild arthritis, struggled to use traditional kitchen tools. He saw an opportunity to create a new line of kitchen tools that were easier to use. Through direct observation, interviews, and testing with users who had similar challenges, the OXO team developed a range of kitchen tools with comfortable, non-slip grips and user-friendly designs. The products were a hit and transformed the kitchenware market, all because the ideation phase was deeply rooted in understanding and solving real user problems.

Proof of Concept – Validating the Vision

The Proof of Concept (PoC) stage stands out as a critical milestone. It's the phase where theoretical ideas confront the realities of physical creation, bridging the gap between imaginative concepts and tangible products. The necessity of PoC in hardware development cannot be overstated; it serves as both a validation of the idea's feasibility and a foundational step in the journey from concept to market-ready product. PoC acts as a litmus test, revealing the viability of an idea under real-world conditions and offering the first glimpse of its potential impact.

The tangible impact of a well-executed PoC is manifold. Firstly, it provides a concrete demonstration of the concept, which is essential not only for internal validation but also for attracting investors and stakeholders. In the hardware realm, where development costs can escalate quickly, a successful PoC can be a convincing tool to secure funding and support.

Setting Objectives for the PoC

Defining Clear Goals: The first step in planning a PoC is to establish what it aims to achieve. This involves setting clear, measurable objectives that focus on validating the core functionalities of the concept. For instance, if you are developing a new type of energy-efficient motor, the PoC should aim to demonstrate its energy-saving potential and basic functionality.

Scope of Demonstration: Decide what aspects of the product will be tested and demonstrated. This doesn't mean building a fully functional prototype; instead, it's about proving the fundamental principle or technology upon which your product is based.

Success Criteria: Establish criteria for what constitutes a successful PoC. This could be technical performance metrics, user experience feedback, or cost-effectiveness.

Resource Allocation: Strategically Channeling Inputs

Time Management: Allocate a realistic timeframe for your PoC. Time is a crucial resource in product development, and it's essential to balance thoroughness with efficiency. For example, setting a timeline that allows for adequate testing and iteration without dragging on too long will be instrumental here.

Budgeting: Determine the budget required to achieve the PoC objectives. This includes costs for materials, labor, and any external services like consulting or specialized testing. Budgeting must be done with foresight, considering potential overruns and unexpected expenses.

Manpower and Skill Allocation: Assign the right people to the PoC. This includes internal team members with the necessary skills and expertise, as well as identifying when external expertise is needed. For instance, if your PoC requires specialized knowledge in a certain type of sensor technology, consider bringing in an expert in that field.

Resource Optimization: It's about using the allocated resources in the most efficient way possible. This could mean leveraging existing technologies or platforms, minimizing waste, or using iterative processes to streamline development.

Risk Assessment in Proof of Concept: Navigating Potential Challenges

The Proof of Concept (PoC) stage, being a crucial early phase in product development, comes with its own set of risks. Understanding these risks and developing effective mitigation strategies are key to ensuring the PoC's success. Let's examine how to identify potential risks and explore an example that highlights the importance of this process.

Identifying Potential Risks in PoC

Technical Feasibility: One of the foremost risks is the technical feasibility of the concept. This includes challenges related to the integration of technologies, material choices, or the complexity of the design.

Resource Limitations: Another significant risk is the misjudgment of the resources required, whether it be time, budget, or expertise. Underestimating these aspects can lead to significant setbacks.

Market Relevance: PoC must also consider the risk of market relevance. Will the concept meet the market needs or consumer expectations? Misalignment with market demand can render a technically successful PoC commercially unviable.

Mitigation Strategies for PoC Risks

Technical Validation: Engage in early-stage technical validation. Use simulations, small-scale experiments, or consult with experts to gauge the technical feasibility of your concept.

Resource Contingency Planning: Have a contingency plan for resources. Allocate a buffer in your budget and timeline to accommodate unforeseen challenges.

Market Analysis: Conduct a preliminary market analysis to ensure that your concept aligns with current and future market trends. This can involve customer surveys, focus groups, or exploring existing market data.

The Segway PoC

An illustrative example of risk assessment in a PoC is the development of the Segway. The Segway, a two-wheeled, self-balancing personal transporter, was a technological marvel when first introduced. However, its development faced several risks:

Technical Risk: The Segway's innovative self-balancing technology was unproven and complex, posing a significant technical risk during its PoC stage. The team mitigated this by extensive testing and iterating the technology to ensure stability and safety.

Market Acceptance Risk: Despite its technical ingenuity, the Segway faced market acceptance risks. Initially, it was unclear how consumers would perceive this novel mode of transportation. The company conducted numerous trials and demonstrations to gauge public reaction and gather feedback, which was crucial in understanding the market's readiness for such a product.

Regulatory Risk: There was also a risk regarding regulatory acceptance, as the Segway didn't fit into traditional transportation categories. Proactively, the company engaged with regulatory bodies to navigate these challenges, ensuring the product met necessary standards and regulations.

Risk assessment in the PoC stage is about foreseeing potential hurdles and preparing to address them effectively. The Segway's journey from a PoC to a marketable product illustrates the importance of identifying and mitigating risks related to technical feasibility, market acceptance, and regulatory compliance.

Technology Scouting in Proof of Concept: Harnessing the Right Tools

In the Proof of Concept (PoC) stage, selecting the right technologies is crucial to validate your idea effectively. This process, known as technology scouting, involves researching, identifying, and selecting technologies that align with the objectives of your PoC.

Identifying Appropriate Technologies

Alignment with PoC Goals: The chosen technologies should directly contribute to achieving the specific objectives of your PoC. This requires a clear understanding of what you aim to demonstrate or test through your PoC.

Research and Exploration: Conduct thorough research to explore available technologies. This can include academic journals, industry publications, technology expos, and discussions with experts in the field. The goal is to stay informed about the latest advancements that could potentially benefit your PoC.

Evaluating Technical Compatibility: Assess how well the shortlisted technologies integrate with each other and with your concept. Compatibility is key to ensuring that different technological components work harmoniously in your PoC.

Selecting the Right Technology Mix

Cost-Effectiveness: Consider the cost implications of the technologies. The aim is to find a balance between technological capability and cost, ensuring that the PoC remains financially viable.

Scalability and Future Integration: While focusing on the immediate needs of the PoC, it's also important to consider how these technologies can scale or integrate into future development phases. However, this should not be confused with the actual development or prototyping of the product.

Ease of Implementation: Evaluate how easy it is to implement and use the selected technologies. User-friendly technology can significantly streamline the PoC process.

Technology Trials and Testing

Experimentation: Before fully integrating technology into your PoC, conduct small-scale trials or experiments. This helps in understanding its practical application and any potential issues that might arise.

Gathering Feedback: Collect feedback from the team working on the PoC about the usability and effectiveness of the technology. This feedback is crucial for making informed decisions about technology adoption.

Iterative Approach: Be prepared to revisit your technology choices based on trial outcomes. An iterative approach allows for flexibility and adaptation, which is essential in the dynamic landscape of technology development.

Technology scouting is a strategic and thoughtful process in the PoC phase, crucial for laying the groundwork for a successful demonstration of your concept. By carefully selecting technologies that align with your PoC's objectives, are cost-effective, and are easy to implement, you set a strong foundation for your idea.

Documentation and Process Management in Proof of Concept

In the Proof of Concept (PoC) phase, effective documentation and process management play a pivotal role. They not only provide a structured approach to capturing the development journey but also ensure that every step, decision, and change is meticulously recorded.

Documentation Strategy: Creating a Comprehensive Record

Purpose of Documentation: Documentation in the PoC stage serves multiple purposes. It acts as a record of the concept's evolution, a communication tool for stakeholders, and a reference for future development phases. It also aids in maintaining transparency and accountability throughout the development process.

Establishing a Documentation Framework: Develop a structured framework for documenting the PoC process. This should include guidelines on what needs to be documented, how it should be recorded, and who is responsible for maintaining these records.

Key Documentation Elements: Essential elements to document include the initial concept description, objectives of the PoC, technical assessments, resource allocations, risk analyses, technology choices, and any changes or iterations made during the process.

Process Management: Ensuring Methodical Development

Standardizing Procedures: Establish standardized procedures for each step in the PoC phase. This includes protocols for conducting assessments, technology trials, and risk mitigation strategies. Standardization ensures consistency and quality in the development process.

Tracking Progress and Changes: Implement a system to track progress and log any changes made during the PoC development. This can be facilitated through project management tools or customized tracking systems. Regular updates and logs help in monitoring the project's trajectory and making informed decisions.

Collaborative Documentation: Encourage a collaborative approach to documentation where all team members contribute. This not only distributes the workload but also ensures diverse perspectives and detailed recording of different aspects of the PoC.

Review and Analysis

Regular Documentation Reviews: Conduct regular reviews of the documentation to ensure accuracy and comprehensiveness. This is also an opportunity to reflect on the development process, identify areas of improvement, and adjust strategies as needed.

Learning from Documentation: The documented records are a rich source of learning. They provide insights into what worked well and what didn't, helping in refining future PoC endeavors and avoiding past mistakes.

Preparing for Future Phases: Documentation from the PoC phase is invaluable for future stages of product development. It provides a detailed history of the concept's inception and development, serving as a foundation for design, prototyping, and further innovation.

Effective documentation and process management are integral to the success of the PoC phase. This meticulous approach is crucial for transforming a concept into a viable prototype and, eventually, a market-ready product, ensuring that every decision and iteration is well-thought-out and purposeful.

Change Management in the PoC Stage:

Effective change management during the Proof of Concept (PoC) stage often necessitates adaptability and responsiveness to new findings, challenges, and evolving objectives.

The Essence of Change Management in PoC

Embracing Flexibility: PoC development is inherently dynamic, with changes often needed to refine the concept or address unforeseen challenges. Effective change management involves embracing these changes as opportunities for improvement rather than obstacles.

Systematic Approach to Changes: Implementing a structured approach to handle changes is crucial. This includes assessing the impact of each change, revising plans, and ensuring that all stakeholders are informed and in agreement with the modifications.

Documentation and Tracking: Documenting changes meticulously is vital. This not only provides a clear record of the evolution of the PoC but also assists in evaluating the efficacy of each change.

Dyson's Cyclone V10 Vacuum Cleaner

Dyson's development of the Cyclone V10 cord-free vacuum cleaner serves as an exemplary case of change management in PoC. The challenge was to create a powerful, yet lightweight and efficient vacuum cleaner.

Iterative Development: The PoC phase involved numerous iterations. Each version brought insights that led to changes in the design, motor efficiency, and battery life. The process was characterized by a willingness to revisit and revise earlier decisions based on new findings and performance data.

Managing Changes: Dyson's team systematically managed these changes. They evaluated the implications of each iteration on the overall design and performance, ensuring that the final product aligned with their vision of a revolutionary vacuum cleaner.

Impact of Changes: Each change brought the team closer to achieving a balance between power and efficiency. The result was a product that was not only innovative but also set new standards in the vacuum cleaner market.

Strategies for Effective Change Management in PoC

Regular Reviews and Feedback Loops: Establish regular review sessions where changes can be proposed, discussed, and decided upon. Feedback loops with team members and stakeholders are crucial for gaining diverse perspectives on proposed changes.

Risk Assessment for Changes: Assess the risks associated with each proposed change. This involves considering the potential impact on the project timeline, costs, and the final product's functionality.

Agility in Decision Making: The ability to make quick yet informed decisions in response to change requests is essential. This agility ensures that the PoC development remains dynamic and responsive to necessary adaptations.

Change management during the PoC phase is about striking a balance between flexibility and structure. It requires an open-minded approach to revisions while maintaining a clear vision of the end goal. Dyson's journey in developing the Cyclone V10 exemplifies how effective management of changes and iterations can lead to breakthrough innovations.

Enhancing Knowledge Capture in Proof of Concept: A Practical Guide

The Proof of Concept (PoC) phase in hardware product development is a goldmine of insights and learnings. Capturing this knowledge systematically is not just about documentation; it's about preserving the essence of innovation and facilitating informed decision-making for future endeavors.

The Vital Role of Knowledge Capture

Scenario Illustration: Imagine developing a PoC for a novel solar-powered device. During testing, certain materials underperform, while others excel unexpectedly. Systematically capturing these findings can inform future material choices, saving time and resources.

Preserving Intellectual Capital: Every test, iteration, and brainstorming session in the PoC phase contributes to the project's intellectual capital. This information, if lost, can mean reinventing the wheel for future teams.

Comprehensive Strategies for Knowledge Capture

Structured Digital Documentation: Use digital tools to maintain detailed records of experiments, design changes, and team meetings. For instance, a digital lab notebook can capture experimental setups and results in real time, providing a valuable reference for future projects.

Interactive Knowledge Sharing: Encourage interactive sessions where team members can share insights and experiences. This could be in the form of weekly round-table discussions, where team members discuss what worked and what didn't, fostering a culture of continuous learning.

Utilizing Multimedia Tools: Diversify documentation methods by incorporating multimedia tools. Short video diaries or audio recordings can capture insights that might be missed in written formats.

Tacit Knowledge and Its Importance

Consider a hypothetical scenario where your team discovers an unconventional method to improve battery efficiency during the PoC phase. This 'tacit knowledge' born from hands-on experience is invaluable and should be recorded through interviews or informal chats, ensuring it is not lost when team members move on.

Creating a Knowledge Repository: Develop an accessible, central knowledge repository. This could be a digital library where all documentation, from formal reports to casual notes, is stored. Regular updates and a user-friendly interface can make this repository a go-to resource for the team.

Capturing Experiential Learning

Reflective Practices: Encourage team members to maintain reflective journals. Here, they can document their daily experiences, challenges faced, and how they overcame them. These personal insights can provide unique perspectives on the PoC process.

Scenario for Experiential Learning: Consider a situation where a particular design approach fails repeatedly. Documenting the steps taken to resolve this issue, and the thought process behind it, can serve as a case study for future problem-solving.

In the PoC phase, effective knowledge capture goes beyond the traditional bounds of documentation. It encompasses a holistic approach that values every piece of information, from hard data to intuitive insights. By implementing structured, creative, and inclusive knowledge capture strategies, teams can build a rich repository of information.

Validating the Proof of Concept: A Deep Dive into Testing Parameters

The validation phase in the Proof of Concept (PoC) process is where the rubber meets the road, determining whether the idea in theory holds up in practice. Establishing robust testing parameters is critical in this phase to accurately assess the PoC's viability.

Defining Testing Parameters

Objective Criteria: The first step is to establish objective, quantifiable criteria that the PoC must meet. These criteria should be directly linked to the core functionalities and objectives of the concept. For instance, if the PoC involves a new type of battery, testing parameters might include charge time, lifespan, and energy output.

Customized Testing Methods: Depending on the nature of the PoC, customize your testing methods. If it's an electronic device, tests might include electrical safety, efficiency, and durability under various conditions. For a mechanical device, stress tests, load capacity, and wear-and-tear analysis could be pertinent.

Ensuring Comprehensive Validation

Simulated Environments: To mimic real-world conditions, conduct tests in simulated environments. This approach helps identify how the concept performs under different scenarios, providing a more holistic view of its effectiveness and reliability.

Iterative Testing: Validation should be an iterative process. Initial tests might reveal areas for improvement, leading to tweaks in the design, followed by re-testing. This cycle continues until the PoC meets all the established criteria satisfactorily.

Addressing Technical and Practical Feasibility

Technical Validation: This involves assessing whether the PoC functions as intended from a technical standpoint. It answers questions like: Does the technology work reliably? Are there any unforeseen technical challenges?

Practical Feasibility: Apart from technical validation, practical feasibility is equally important. This includes assessing the ease of manufacturing, cost-effectiveness, and user-friendliness of the concept.

Documentation and Analysis

Recording Results: Document all testing results meticulously. This documentation should include not just the outcomes but also the methodologies used, conditions of each test, and any anomalies observed.

Data-Driven Decisions: Use the results from these tests to make informed decisions about the next steps. If the PoC passes validation, it's ready for further development. If not, the data provides crucial insights into what needs reworking.

Validating a PoC is a critical step in the journey of bringing a hardware concept to life. Establishing and executing well-thought-out testing parameters ensures that the concept is not only theoretically sound but also practically viable.

Proof of Concept: Strategic Stakeholder Engagement

Masterful communication and engagement with stakeholders during the Proof of Concept (PoC) stage are pivotal for hardware product developers. This stage is not just about showcasing the PoC but also about building a narrative that resonates with different stakeholders, from investors to potential users.

Stakeholder Analysis

Identifying and Categorizing Stakeholders: Begin by mapping out all potential stakeholders, categorizing them based on their influence and interest. This could range from venture capitalists and angel investors to industry experts, potential customers, and supply chain partners.

Understanding Stakeholder Dynamics: Analyze the motivations, expectations, and concerns of each group. Investors might prioritize financial projections and market scalability, while end-users might be more focused on usability and practicality.

Tailoring the Communication Strategy

Crafting a Compelling Narrative: Develop a storytelling approach that connects the PoC's value proposition to each stakeholder's interests. For investors, highlight the innovation's potential market disruption and return on investment. For users, focus on how the product solves a key problem or improves their quality of life.

Clear and Engaging Messaging: Use language and terminology that is accessible and engaging. Avoid technical jargon that might alienate non-technical stakeholders. Aim to evoke a sense of excitement and possibility about the PoC's potential.

Effective Presentation Approaches

Dynamic Demonstrations: If feasible, arrange live demonstrations of the PoC. Allowing stakeholders to interact with the prototype can create a more lasting impression than abstract descriptions.

Utilizing Multimedia Tools: Leverage high-quality visuals, animations, or interactive digital models to explain complex technical concepts. This approach can help stakeholders visualize the end product and its applications.

Interactive Feedback Sessions

Encouraging Open Dialogue: Create an environment where stakeholders feel comfortable voicing their opinions. Be open to constructive criticism and use these sessions as an opportunity to gather diverse perspectives.

Adaptive Response Mechanism: Prepare to address queries and concerns on the spot. Show that you value their input and are committed to incorporating feasible suggestions into future iterations.

Building and Maintaining Relationships

Long-Term Relationship Building: View stakeholder engagement as an ongoing relationship rather than a one-off interaction. Regular updates on the PoC's progress can keep stakeholders involved and invested in the project's success.

Post-Presentation Engagement: Follow up with personalized communications, addressing specific interests or concerns raised during the presentation. This follow-up can reinforce their importance to the project and keep the conversation going.

Effective stakeholder engagement in the PoC stage is about much more than just presenting an idea; it's about creating a shared vision for the future. By thoroughly understanding stakeholder needs, tailoring communication strategies, using dynamic presentation techniques, fostering interactive feedback, and building enduring relationships, hardware product developers can set a solid foundation for their project's success.

The Art of Storytelling in PoC Presentation

Narrative development in the context of presenting a Proof of Concept (PoC) is a crucial skill, enabling hardware product developers to effectively communicate their vision and the potential of their product. A well-crafted narrative conveys the technical aspects of the PoC and weaves a story that resonates emotionally with stakeholders. Let's explore how to create a compelling narrative for the PoC.

Establishing the Core Message

Defining the Essence of the PoC: Start by identifying the core message of your PoC. What is the primary problem it addresses? How does it innovate or improve upon existing solutions? This core message will form the backbone of your narrative.

Highlighting Unique Value Proposition: Emphasize what sets your PoC apart from others. Whether it's a novel application, exceptional efficiency, or cost-effectiveness, make sure this unique selling point is front and center in your story.

Structuring the Narrative

Creating a Story Arc: Structure your narrative in a way that captivates and maintains interest. Begin with an introduction that sets the context, followed by a description of the problem or need, the journey of developing the PoC, and culminating with the potential impact and future possibilities.

Incorporating Emotional Elements: Engage your audience emotionally by sharing challenges faced and overcome, moments of insight, or the inspiration behind the PoC. This humanizes the technical aspects and makes the story relatable.

Connecting with the Audience

Understanding Stakeholder Perspectives: Tailor your narrative to resonate with the specific interests and concerns of your audience.

What matters to an investor might be different from what intrigues a potential partner or end-user.

Interactive Storytelling: Encourage questions and interactions during your presentation. This not only keeps the audience engaged but also allows you to adjust your narrative in real time based on their responses.

Concluding with a Call to Action

Presenting a Clear Call to Action: Conclude your narrative with a clear and compelling call to action. Whether it's seeking investment, partnership, or feedback, make sure your audience knows what you expect from them next.

Reiterating the Vision: End by reiterating the long-term vision for your PoC. Leave your audience with a lasting impression of the potential impact and future possibilities of your product.

Crafting a compelling narrative around your PoC entails telling a story that inspires, engages, and prompts action. Remember, a good story not only informs but also transforms the way people perceive your PoC.

Case Study: The Tale of Ring Video Doorbell's PoC Success

In examining successful Proof of Concept (PoC) projects, it's enlightening to explore real-life examples that have significantly impacted their respective industries. One such instance is the story of the Ring Video Doorbell, a product that revolutionized home security and convenience. This case study not only demonstrates the tangible impact of a well-executed PoC but also offers valuable insights into the journey from a simple idea to a market-changing product.

The Genesis of Ring Video Doorbell

The concept behind Ring was born from a desire to enhance home security in a user-friendly way. The founder, Jamie Siminoff, identified a gap in the market: the need for a more interactive and accessible approach to home security. The initial idea was simple yet innovative—a doorbell equipped with a video camera that allowed homeowners to see and speak with visitors via their smartphones, regardless of their physical location.

Developing the PoC

The PoC for Ring was crucial in validating the concept's feasibility and practicality. The initial prototype was rudimentary, cobbled together from various off-the-shelf components. Despite its simplicity, this early version was pivotal in demonstrating the core functionality of the product: real-time video communication.

Challenges and Breakthroughs

The development of Ring's PoC wasn't without challenges. One of the major hurdles was integrating different technologies—video cameras, motion sensors, and mobile connectivity—into a seamless user experience. The PoC phase involved rigorous testing and iterations to refine these integrations. There was also the challenge of designing a product that was not only functional but aesthetically pleasing and easy to install.

Stakeholder Engagement and Feedback

Throughout the PoC development, engaging with potential users and stakeholders was a key strategy. Early demonstrations to friends, family, and potential investors provided valuable feedback, which was instrumental in refining the concept. This feedback loop helped in identifying and addressing user concerns, such as ease of installation and use, leading to significant improvements in the design.

Market Impact and Success

The PoC's success paved the way for further development and eventual market launch. Once launched, Ring Video Doorbell quickly gained popularity, heralded for its innovative approach to home security. It offered a unique combination of convenience, security, and peace of mind, resonating with a broad consumer base. Ring's success attracted significant investment, and eventually, the company was acquired by Amazon, further cementing its position in the market.

Key Insights from Ring's PoC Journey

Addressing a Genuine Need: Ring's success underscores the importance of developing a product that addresses a real market need. Understanding and solving a specific problem can be the key to a product's relevance and appeal.

Iterative Development: The iterative nature of Ring's PoC highlights the importance of continuous refinement based on real-world testing and feedback.

Stakeholder Engagement: Active engagement with potential users and stakeholders during the PoC phase can provide critical insights that shape the product's development.

Integration of Technologies: Ring demonstrated how integrating various technologies could create a novel product, emphasizing the need for seamless technological synergy in product development.

The journey of Ring Video Doorbell from a PoC to a market-leading product is a testament to the power of a well-executed Proof of Concept. It showcases how a simple idea, when developed thoughtfully through the PoC stage, can lead to a product that not only succeeds commercially but also makes a meaningful impact in people's lives.

Preparing for Further Development: Navigating from PoC to Detailed Design

The transition from the Proof of Concept (PoC) phase to the detailed design phase is a critical juncture in the hardware product development cycle. This stage marks the evolution of an idea from a validated concept to a blueprint for a tangible product. Understanding how to effectively navigate this transition is key to maintaining momentum and ensuring the viability of the final product.

Outlining the Transition Steps

Review and Analysis of PoC Outcomes: The first step is to conduct a comprehensive review of the PoC outcomes. This involves analyzing the data, feedback, and learnings obtained from the PoC phase. Key areas of focus include the PoC's technical performance, user feedback, and any identified shortcomings or areas for improvement.

Refinement of Product Concept: Based on the PoC outcomes, refine the initial product concept. This may involve modifying the product's features, design, or functionalities to better align with user needs, technical feasibility, and market demands.

Detailed Design Planning: Develop a detailed plan for the design phase, outlining the specific tasks, timelines, and resource requirements. This plan should consider all aspects of product development, including industrial design, engineering, materials selection, and user experience.

Integration of Stakeholder Feedback: Incorporate feedback from key stakeholders, including potential customers, partners, and team members. This step ensures that the product design is aligned with market expectations and addresses any concerns or suggestions raised during the PoC phase.

Technical Specification Development: Create a detailed technical specification document that outlines all technical requirements and design parameters for the product. This document serves as a blueprint for the design and development team and should include specifications for materials, components, functionality, performance, and compliance requirements.

Risk Assessment and Mitigation for Design Phase: Identify potential risks and challenges that may arise during the design phase and develop strategies to mitigate these risks. This may include technical challenges, budget constraints, timeline delays, or market shifts.

Allocation of Resources: Allocate the necessary resources, including manpower, budget, and technology, to support the design phase. Ensure that the team has the skills, tools, and support needed to execute the design plan effectively.

Establishing Protocols for Iterative Design: Set up a process for iterative design and testing. This process should allow for continuous refinement of the design based on ongoing testing, feedback, and technical advancements.

Preparation for Regulatory Compliance: Begin preparations for meeting regulatory compliance and industry standards during the design phase. This ensures that the product design considers all necessary compliance requirements from the outset.

Documentation and Knowledge Transfer: Ensure thorough documentation of the transition process and the rationale behind key design decisions. This documentation is crucial for future reference and for facilitating a smooth knowledge transfer to new team members or stakeholders.

Transitioning from the PoC to the detailed design phase is a complex but vital process that sets the foundation for the product's future success. It requires careful planning, stakeholder alignment, and a strategic approach to risk management and resource allocation.

Chapter 06

Design – Transforming Vision into Reality

The design phase in hardware product development is a transformative journey where a Proof of Concept (PoC) evolves into a meticulously crafted design. This stage is crucial for aligning a product with its intended vision and market needs. A compelling example of this transition is the development of the Pebble Smartwatch, a pioneering product in the wearable technology space.

The Pebble Smartwatch: Pioneering the Wearable Tech Design

The Pebble Smartwatch story began as a simple idea: to create a watch that could display messages from a smartphone. This concept, initially a PoC, navigated through an intense design phase that addressed both technological innovation and consumer expectations. The Pebble team focused on creating a design that was functional, user-friendly and aesthetically pleasing.

Overview of Design Objectives in Pebble's Journey

Aligning with the Vision: The Pebble's design was driven by the vision to seamlessly integrate the utility of a smartphone with the convenience of a wristwatch. This meant designing a product that was compact yet capable of handling multiple functionalities like notifications, fitness tracking, and customizable watch faces.

Responding to Market Needs: Understanding and integrating market needs were critical. The Pebble design included features like an e-paper display for longer battery life and better outdoor readability, addressing a key consumer demand for wearable technology that was practical for everyday use.

Balancing Innovation and Practicality: The design phase for Pebble was marked by a balance between innovative features and practical manufacturing. The team faced challenges in integrating components like a small yet powerful battery, a readable display, and a water-resistant casing, all within a design that was comfortable and appealing to wear.

Incorporating User Feedback: User feedback was integral to Pebble's design process. Early prototypes and Kickstarter campaigns provided invaluable insights into user preferences, leading to design tweaks such as the inclusion of physical buttons for easier navigation, which users preferred over a touchscreen interface.

Designing for a Niche Market: Pebble's design strategy also involved focusing on a niche market of tech enthusiasts and early adopters. This focus allowed the team to create a design that catered to specific user needs, setting the stage for broader market acceptance.

The Pebble Smartwatch's journey from a PoC to a groundbreaking product in wearable technology exemplifies the critical role of the design phase in hardware product development. It highlights the importance of aligning design with product vision, responding to market demands, and striking a balance between innovation and manufacturability.

Developing the Specification Requirement Document (SRD)

Purpose and Importance

The Specification Requirement Document is the foundational blueprint that turns a product concept into a tangible reality. It's the guiding star for the entire design process, laying down the specifications and expectations in clear, measurable terms.

Communication Bridge: The SRD acts as a pivotal communication tool, ensuring that all teams – from design and engineering to marketing and manufacturing – are aligned in their understanding and objectives. It's like the playbook of a sports team where every member knows their role and the overall game strategy.

Strategic Decision-Making Guide: The SRD is essential for informed decision-making. It helps teams evaluate various design options, weighing factors like technical feasibility, user needs, and cost implications. This strategic tool prevents deviation from core objectives while allowing flexibility in approach.

Risk Mitigation Framework: Identifying potential design risks early is crucial. The SRD allows for a preemptive strategy to manage these risks, reducing the likelihood of costly and time-consuming revisions down the line. It's equivalent to having a contingency plan in place for unforeseen challenges.

Benchmark for Quality: This document is the yardstick against which the final product is measured. It serves as a constant reminder of the quality standards and specifications that the product must adhere to, ensuring that the vision for the product is accurately realized.

Cost Management Template: Managing costs is integral to product development. The SRD provides a detailed breakdown of materials, components, and processes, allowing for precise cost estimation and control. It helps in identifying cost-saving opportunities without compromising on product integrity.

Key Components of a Robust SRD

An effective SRD is a meticulous and comprehensive document covering several critical aspects:

Functionality Descriptions: This section is the core of the SRD, detailing the product's intended functions and features. It's not just a list but a narrative explaining how each feature adds value to the user. For instance, in a smart home device, this could include specific functionalities like remote control via an app, energy-saving modes, or compatibility with other smart devices.

Detailed Performance Specifications: Performance specs are the quantifiable goals of the product, such as speed, endurance, capacity, and efficiency. For example, if designing a battery-powered device, this would specify battery life under various usage scenarios and charging time.

Physical and Environmental Constraints: This covers the physical limitations, such as dimensions, weight, and materials, and environmental conditions like temperature range, humidity tolerance, and exposure to elements. These specifications are crucial for ensuring that the product is practical and durable in its intended setting.

User Interface and Experience (UI/UX) Details: This involves a deep dive into how users will interact with the product. It includes the ergonomic design of controls, the intuitiveness of the interface, and the aesthetics of the product. This section is about creating a user-centric design that is both functional and appealing.

Comprehensive Regulatory Compliance: This critical component ensures the product meets all relevant industry standards and regulations, which can vary significantly depending on the market and product type. It involves detailed research into legal requirements, safety standards, and certification processes.

Materials and Sustainability Considerations: This section delves into the materials selected for the product, focusing on factors like durability, cost-effectiveness, and environmental impact. It's about making informed choices that balance product requirements with sustainability goals. For instance, selecting biodegradable materials for packaging or using recycled plastics for certain components can be pivotal decisions.

Manufacturability and Assembly Insights: Here, the focus is on the practical aspects of producing and assembling the product. This includes considerations like ease of assembly, cost-effectiveness of production methods, and scalability. This part of the SRD is crucial for bridging the gap between design and manufacturing, ensuring that the product is not only well-designed but also feasibly manufacturable at scale.

Integration of Feedback Mechanisms: A forward-thinking SRD also plans for user feedback integration. This involves setting up channels and processes for gathering user feedback post-launch and how this feedback will be used to inform future design iterations.

In essence, the SRD encompasses everything from detailed functionality, performance standards, and physical constraints to UI/UX design, regulatory compliance, and cost analysis. By providing a clear and thorough outline of all these aspects, the SRD ensures that the final product not only meets but exceeds market and user expectations.

Additionally, the SRD is not a static document; it's dynamic and should evolve as the product moves through different stages of development. This adaptability is key to responding to new insights, technological advancements, and changing market trends. A well-maintained SRD serves as a living document that reflects the current state of the product design and guides future modifications.

To make the SRD as effective as possible, it's important to involve a diverse team in its creation and upkeep. Input from various departments – including design, engineering, marketing, and customer service – ensures that the document covers all bases and that the product is developed with a holistic view.

In crafting the SRD, clarity and precision are paramount. The document should be detailed enough to provide clear guidance but also flexible enough to allow for creative solutions. It's a delicate balance between providing enough information to guide the design process and leaving room for innovation and improvement.

Design for Manufacturability (DFM)

In product design, Design for Manufacturability (DFM) is a strategic approach that focuses on designing products with ease of manufacturing in mind. This concept is vital for reducing production costs, enhancing product quality, and ensuring efficient manufacturing processes.

Principles of DFM

Simplification of Design: The primary goal of DFM is to streamline the product design, making it as simple as possible. This includes reducing the number of parts, standardizing components, and eliminating unnecessary complexities, which can lead to easier assembly and fewer errors.

Utilization of Standard Components: Leveraging standard, off-the-shelf components ensures greater availability, cost-effectiveness, and reliability. Standard parts often come with well-understood specifications and manufacturing processes.

Ease of Assembly: A key aspect of DFM is designing products that are straightforward to assemble. This involves creating parts that are easy to handle and assemble, minimizing the use of different types of fasteners, and designing for automation where possible.

Early Integration of Manufacturing Constraints: It's critical to consider manufacturing limitations early in the design process. This includes the capabilities and constraints of available machinery, tooling, and technologies.

Material and Process Optimization: Choosing the right materials and manufacturing processes for each component is crucial. This decision should consider factors like cost, functionality, durability, and ease of manufacture.

OXO Good Grips Kitchen Tools

An insightful example of a successful DFM application can be seen in OXO's Good Grips line of kitchen tools. OXO, a brand that focuses on ergonomic design, applied DFM principles effectively in creating their innovative kitchen utensils.

Design Simplification: OXO's Good Grips tools are designed with simplicity and functionality in mind. For instance, their handles are made with a simple, ergonomic design that not only provides comfort but also makes the manufacturing process more straightforward.

Standardization of Components: OXO utilized standard components across various products. For example, the signature non-slip, black rubber handle is a common feature in many of their tools, which simplifies manufacturing and reduces costs.

Ease of Assembly and Use: The products are designed to be easily assembled, with fewer parts and a focus on intuitive construction. This makes the manufacturing process more efficient and enhances the user experience.

Material Selection for Manufacturability: OXO carefully selected materials that are durable, functional and suitable for efficient manufacturing processes. The use of stainless steel and high-quality rubber in their products is a testament to this approach.

Incorporation of User Feedback: An integral part of OXO's design philosophy involves incorporating user feedback, which aligns with DFM by ensuring that the products are not only manufacturable but also meet the needs and preferences of the users. This feedback loop helps OXO to continuously refine their products, making them easier to produce and more user-friendly.

Manufacturing Process Optimization: OXO consistently works to optimize their manufacturing processes in line with their product designs. This includes using efficient molding techniques for their rubber handles and precision machining for the metal parts, ensuring high-quality finishes and consistent product quality.

Balancing Cost and Quality: OXO's application of DFM principles extends to balancing the cost of production with the quality of the product. By optimizing the design and manufacturing process, they manage to keep the products affordable without compromising on quality or functionality.

Focus on Sustainability: In recent years, OXO has also started to incorporate sustainability into their DFM approach. This includes selecting materials that are not only easy to manufacture but also environmentally friendly, contributing to a more sustainable product lifecycle.

OXO's Good Grips kitchen tools exemplify a successful implementation of Design for Manufacturability.

Benefits and Challenges: Balancing Design Complexity with Manufacturability

Design for Manufacturability (DFM) is a strategic approach in product development, emphasizing the design of products for ease of production. This methodology brings significant advantages but also presents some challenges, necessitating a fine balance between the intricacies of design and practical manufacturing considerations.

Advantages of DFM

Cost Efficiency: DFM aids in reducing production expenses by streamlining designs and utilizing standard components. Simplified designs often necessitate fewer materials and labor, leading to considerable cost savings.

Enhanced Product Quality and Uniformity: Designs that are simple and standardized tend to be more reliable, leading to a reduction in defects and an elevation in product quality.

Improved Manufacturing Productivity: Designs that are straightforward to produce increase the efficiency of manufacturing processes, reducing the time required to bring products to market.

Flexibility in Manufacturing: DFM can lead to designs that are adaptable, allowing for quick adjustments in response to market shifts or technological advancements.

Challenges of DFM

Design Limitations: A significant challenge lies in adhering to manufacturing constraints without sacrificing the product's functionality or aesthetic qualities.

Interdepartmental Collaboration: Effective DFM necessitates close cooperation between design and manufacturing teams, which can be complex, especially in larger organizations.

Equilibrium Between Cost and Quality: While minimizing costs is a primary aim of DFM, it is essential to ensure that this does not result in a decline in product quality.

Technological Evolution: Keeping abreast of advancements in manufacturing technology requires ongoing learning and adaptation.

Case Study: LEGO Bricks

An exemplary case of DFM in action is seen in the manufacturing of LEGO bricks, a product that exemplifies the balance between design intricacy and manufacturability.

Design Precision and Elegance: Each LEGO brick is crafted with meticulous precision, allowing for seamless interlocking. The design is straightforward yet versatile, facilitating a vast array of construction possibilities.

Uniformity and Interoperability: LEGO's significant achievement lies in the uniformity of its components. Despite the diversity in shapes and sizes, LEGO bricks are compatible and can be combined in innumerable ways. This uniformity simplifies the manufacturing process and enhances the consumer experience.

Efficient Material Selection: LEGO utilizes acrylonitrile butadiene styrene (ABS), a thermoplastic polymer, for most of its bricks. ABS is notable for its durability, consistency in melting point, and suitability for high-volume injection molding, making it an optimal material choice for efficient manufacturing.

Refinement of Manufacturing Techniques: Over the years, LEGO has continuously refined its manufacturing techniques to ensure precision and efficiency. The injection molding process used is highly effective, producing bricks with exceptional accuracy, crucial for ensuring the perfect fit of each piece.

Adaptability to Consumer Demands: LEGO has adeptly adapted its product range to align with evolving consumer preferences while adhering to its core DFM principles. This includes introducing innovative colors, shapes, and themed sets, all compatible with the traditional LEGO system.

Stringent Quality Assurance: LEGO's commitment to quality is evident in its meticulous manufacturing process. Each brick undergoes extensive quality checks, ensuring adherence to the company's stringent standards. This commitment not only reinforces consumer trust but also exemplifies the success of their DFM strategy in maintaining high product standards.

Integration of Environmental Responsibility: In recent years, LEGO has incorporated sustainability into its DFM strategy, exploring eco-friendly materials that comply with their manufacturing standards, and showcasing how DFM principles can coexist with environmental stewardship.

Navigating Challenges and Evolving: The path of LEGO in the toy industry has involved navigating various challenges. Continuously innovating within the DFM framework, while maintaining the integrity of its core product, illustrates LEGO's proficiency in applying DFM principles effectively.

Influence on Industry Standards: LEGO's rigorous application of DFM principles has established a benchmark in the toy industry. Their strategy demonstrates how a well-implemented DFM plan can lead to a product that is not only efficiently manufacturable but also widely esteemed for its quality and versatility.

LEGO bricks serve as an illustrative example of the effective application of Design for Manufacturability. LEGO's journey in embracing and innovating within the DFM framework provides insightful lessons for product developers seeking to understand the tangible application of these principles in real-world scenarios.

Design to Cost

Design to Cost is an essential strategy in product development, focusing on controlling and understanding the elements that influence the cost associated with the design of a product. This methodology ensures that the final design adheres to both performance specifications and budget constraints.

Comprehending Cost Drivers in Design

Material Choices: The selection of materials significantly influences design costs. Each material varies in price, availability, durability, and manufacturing requirements. Opting for more affordable materials that fulfill the product's functional and aesthetic needs can markedly decrease overall expenses.

Design Intricacy: The intricacy of a product's design directly affects manufacturing expenses. Complex designs may necessitate sophisticated manufacturing techniques, specialized equipment, or increased labor, all contributing to higher costs. Simplifying the design, while retaining functionality, can result in substantial savings.

Manufacturing Methodologies: The methods chosen for manufacturing have a considerable impact on costs. Labor-intensive processes or those requiring costly machinery escalate expenses. Design teams must find a balance between the ideal manufacturing method and cost considerations.

Standardization and Modular Design: Employing standardized components and modular designs can lead to cost reductions. Standard parts are often more available and economical than bespoke ones. Modular designs promote efficient material usage and streamlined manufacturing.

Production Scale Economies: Design decisions should account for production scale. Designing for large-scale production can capitalize on economies of scale, reducing the cost per unit as production quantities increase. This approach may require initial investments in tooling but leads to lower costs per unit over time.

Tooling and Initial Setup Expenditures: The expenses related to tooling and initial manufacturing setup can be considerable. Designs that necessitate unique or intricate tooling increase costs. Minimizing the need for specialized tooling can lead to notable cost reductions.

Supply Chain and Distribution Management: The cost of procuring components and materials, along with the logistics of transport to the manufacturing site, influences the overall design cost. Effective management of the supply chain and logistics can help decrease these expenses.

Energy Consumption in Manufacturing: The energy required for production can be an overlooked cost factor. Designs that enable energy-efficient manufacturing can significantly reduce costs over time.

Longevity and Maintenance Considerations: Creating products that are durable and straightforward to maintain can lower long-term costs. This approach might involve higher initial expenses but can result in savings through decreased warranty claims and customer service demands.

Compliance and Safety Standards: Meeting regulatory compliance and safety standards can affect design costs. Incorporating these requirements early in the design process is vital to avoid expensive redesigns or modifications later.

A deep understanding and effective management of various cost influencers during the design phase are key to successfully implementing the Design to Cost approach. The essence of Design to Cost lies in its ability to balance creative and technical aspirations with pragmatic financial considerations. Design teams must navigate this delicate balance, making informed decisions that optimize costs without compromising the core value and appeal of the product. This approach demands a thorough understanding of the entire product lifecycle, from raw materials and production processes to maintenance and end-of-life disposal.

Effective application of Design to Cost involves continuous evaluation and iteration throughout the design process. It is not a one-time assessment but a dynamic practice that adapts as designs evolve and new information emerges. This iterative process ensures that cost considerations are integrated into every stage of design, leading to a product that is cost-effective and aligned with market expectations and consumer needs.

Case Study: IKEA Furniture

IKEA, the Swedish furniture company, exemplifies effective strategies for cost-effective design while maintaining quality. Their approach to design centers around several key strategies:

Flat-Pack Design: IKEA's furniture is designed to be flat-packed, reducing shipping

and storage costs significantly. This innovative packaging design allows for more efficient transportation, leading to lower costs for both the company and the customers.

Efficient Material Usage: IKEA often uses particleboard in its furniture, which is less expensive than solid wood but still provides durability and aesthetic appeal. This choice of material balances cost with quality, catering to budget-conscious consumers without compromising the product's longevity.

Modular and Standardized Components: Many IKEA products feature modular designs with standardized components. This standardization allows for large-scale production, reducing manufacturing costs. The modular nature also offers customers flexibility, as they can mix and match different parts according to their needs.

Design for Assembly: IKEA's furniture is designed for easy assembly by the consumer. This self-assembly model reduces the costs associated with pre-assembled furniture, including labor and transportation expenses.

Sustainability Initiatives: IKEA has increasingly focused on sustainability, which is not only an ethical choice but also a cost-saving strategy in the long run. Using renewable materials and focusing on recyclable designs aligns with modern consumer preferences and reduces waste management costs.

IKEA's success in balancing cost and quality demonstrates how strategic design choices can lead to products that are affordable, durable, functional, and appealing to consumers. Their approach offers a valuable lesson in cost-effective design, showing how thoughtful decisions at the design stage can have far-reaching impacts on both costs and customer satisfaction.

Cost-effective design is about making informed, strategic choices that optimize resources and add value. By carefully considering materials, simplifying designs, adopting modern manufacturing methods, and prioritizing key features, designers can create products that strike the perfect balance between affordability and quality. IKEA's model is a testament to the success of this approach, showcasing how innovation and efficiency in design can lead to products that resonate with consumers both in terms of quality and price.

Precompliance: Navigating Regulatory Requirements

Precompliance, in the context of product design, is a critical phase where the design is scrutinized to ensure it meets relevant industry standards and regulatory requirements. This process is vital for any product that will enter the market, as it significantly influences design decisions and can impact the overall success and legality of the product.

Importance of Precompliance

Ensuring Legal Conformity: Precompliance is essential to ensure that the product design adheres to all relevant laws and regulations. Non-compliance can lead to legal issues, including fines and product recalls, which can be costly and damaging to the brand's reputation.

Safety and Quality Assurance: By meeting regulatory standards, compliance helps in assuring the safety and quality of the product. This is particularly crucial in industries like healthcare, automotive, and children's products, where safety is paramount.

Market Access and Acceptability: Compliance with standards is often a prerequisite for entering certain markets. Meeting these requirements is therefore crucial for broad market access and acceptability of the product.

Risk Mitigation: Precompliance helps in identifying and mitigating risks early in the design process. Addressing potential compliance issues during design rather than after production saves time and resources and reduces the risk of costly redesigns.

Steps for Achieving Pre-compliance

Research and Understanding Regulations: The first step is to thoroughly research and understand the regulatory requirements relevant to the product and its market. This involves staying updated with current standards and regulations, which can vary significantly depending on the industry and geographic location.

Incorporating Regulatory Requirements into Design: Once the requirements are understood, they must be integrated into the product design from the outset. This may involve specific materials, design specifications, safety features, or performance standards.

Consulting with Regulatory Experts: Engaging with regulatory experts or consultants can be invaluable in navigating complex regulatory landscapes. They can provide insights and guidance on meeting compliance standards and often have up-to-date knowledge of changing regulations.

Pre-compliance Testing: Conducting pre-compliance testing during the design phase is crucial. This involves testing prototypes or models of the product against the identified regulatory standards to ensure adherence. These tests might include safety, performance, durability, and environmental impact assessments.

Documentation and Record Keeping: Maintaining detailed records of compliance efforts is essential. Documentation should include design specifications, test results, certifications, and any consultations with regulatory experts. This documentation is vital for proving compliance during audits and inspections.

Continuous Monitoring and Adaptation: Regulatory standards can change, so it's important to continuously monitor for any updates and adapt the product design accordingly. This proactive approach ensures ongoing compliance and can prevent future legal and operational challenges.

Collaboration with Manufacturing and Supply Chain Partners: Ensuring that manufacturing processes and supply chain partners also comply with relevant regulations is a critical component of precompliance. Collaboration and communication with these partners are necessary to maintain compliance throughout the production process.

Precompliance is a vital aspect of the design process, ensuring that products meet necessary regulatory standards and industry requirements. By integrating compliance into the design from the beginning, companies can mitigate risks, ensure safety and quality, and secure market access for their products. This process requires ongoing vigilance, collaboration, and a commitment to adhering to regulatory standards, which are essential for the success and longevity of the product in the market. The steps to achieve precompliance involve a detailed understanding of regulations, integration of these requirements into the design, expert consultations, rigorous testing, thorough documentation, and continuous adaptation to changing standards.

Design Failure Mode and Effects Analysis (DFMEA)

Design Failure Mode and Effects Analysis, commonly known as DFMEA, is a structured approach used in product design to identify and mitigate potential failures. The essence of DFMEA lies in preemptively recognizing areas where a product's design might fail and understanding the potential impacts of such failures, thereby enhancing the reliability and safety of the final product.

Concept and Application of DFMEA

Identification of Potential Failure Modes: The primary step in DFMEA is to systematically identify all conceivable ways in which each component of the design could fail. This process involves a detailed analysis of every part, subsystem, and interface in the product design to uncover potential weaknesses.

Assessment of Failure Effects: Once potential failure modes are identified, the next step is to assess the effects of these failures. This includes understanding how the failure of a particular component could impact the overall functionality of the product, the severity of these effects on the end-user, and the possible implications for safety and compliance.

Determination of Failure Causes: After identifying the failure modes and their effects, DFMEA involves investigating the root causes of these potential failures. This step requires an in-depth analysis of design materials, processes, and environmental factors that could contribute to each identified failure mode.

Risk Priority Number (RPN) Calculation: DFMEA employs a Risk Priority Number system to prioritize the identified risks. The RPN is calculated by multiplying the severity of the effect, the likelihood of occurrence, and the detectability of the failure mode. This quantification helps in prioritizing which failure modes need immediate attention and resources.

Development of Mitigation Strategies: Based on the RPN, the design team develops strategies to mitigate high-priority risks. This might involve redesigning certain components, using different materials, enhancing quality control measures, or incorporating additional safety features.

Implementation and Monitoring: The proposed mitigation strategies are then implemented into the design. Continuous monitoring is essential to ensure that these modifications effectively reduce the risk and do not introduce new failure modes.

Documentation and Communication: Maintaining detailed records of the DFMEA process is crucial. This documentation should include the identified failure modes, their effects and causes, RPN scores, and the mitigation strategies adopted. Effective communication of these findings and actions with all relevant stakeholders is essential for collaborative risk management.

Iterative Process: DFMEA is not a one-time activity but an iterative process that continues throughout the product development lifecycle. As designs evolve and new data become available, the DFMEA should be revisited and updated to reflect the current state of the product.

Integration with Other Quality Processes: DFMEA is often integrated with other quality and reliability processes, such as Quality Function Deployment (QFD) and Statistical Process Control (SPC), to provide a comprehensive approach to product quality and reliability.

DFMEA is an indispensable tool in the product design process, providing a systematic approach to identifying and mitigating potential design failures. Its proactive nature helps in enhancing the safety, reliability, and quality of products, ultimately contributing to customer satisfaction and product success.

Conducting DFMEA: A Step-by-Step Process

Conducting a DFMEA involves several critical steps, each aimed at ensuring the design is as robust, safe, and reliable as possible.

Step 1: Review the Design

- Begin with a thorough review of the product design. This includes detailed analysis of drawings, specifications, and any existing prototypes.

- Understand the function of each component, how they interact within the system, and the intended use of the final product.

Step 2: Assemble the DFMEA Team

- Form a cross-functional team that includes members from design, engineering, manufacturing, quality, and any other relevant department.

- The diverse perspectives and expertise of this team are crucial for identifying potential failure modes that might not be apparent to a single department.

Step 3: Identify Potential Failure Modes

- Systematically list potential ways each component or subsystem could fail. Consider mechanical failures, electrical issues, material degradation, software errors, and user-related failures.

- This step requires a balance of technical knowledge and creative thinking to anticipate less obvious failure modes.

Step 4: Determine Failure Effects

- For each identified failure mode, assess the potential consequences. This includes the effect on the product's performance, safety implications, and the potential impact on the end-user.

- Severity of each failure effect should be evaluated, often on a scale from minor inconvenience to critical safety hazard.

Step 5: Identify Causes of Failure

- Analyze the root causes for each potential failure mode. Causes could range from design flaws and material weaknesses to manufacturing variances and environmental factors.

- Understanding the causes is key to developing effective mitigation strategies.

Step 6: Calculate Risk Priority Number (RPN)

- Assign a severity rating (usually 1-10) to each failure mode based on its potential impact.

- Estimate the likelihood of occurrence for each failure mode, again using a scale (1-10).

- Assess the detectability of each failure mode, rating how likely it is that the failure would be detected before reaching the end-user.

- Multiply these three numbers (severity, occurrence, detectability) to get the Risk Priority Number for each failure mode. Higher RPNs indicate higher priorities for mitigation.

Step 7: Develop Mitigation Strategies

- For failure modes with high RPNs, develop strategies to reduce either the severity, the likelihood of occurrence, or improve detectability.

- Strategies might include design changes, material substitutions, additional testing, or enhanced quality control measures.

Step 8: Implement and Monitor Changes

- Implement the identified mitigation strategies in the design.

- Continuously monitor the effectiveness of these changes, ensuring that they effectively reduce the risk without introducing new issues.

Step 9: Document and Update the DFMEA

- Meticulously document the entire DFMEA process. This documentation should include the identified failure modes, their effects and causes, the RPN scores, and the mitigation strategies implemented.

- Ensure that this documentation is accessible and updated regularly. It serves as a vital record for understanding the design decisions made and provides a reference for future design modifications or similar projects.

Step 10: Continuous Review and Improvement

- Recognize that DFMEA is not a one-time task but an ongoing process. As the design evolves, as new materials or technologies become available, or as manufacturing processes change, the DFMEA should be reviewed and updated.

- Encourage regular DFMEA reviews as part of the design lifecycle, particularly at major milestones or after significant changes.

Step 11: Integrate with Other Processes

- For comprehensive risk management, integrate the findings and actions from DFMEA with other design and quality processes like Quality Function Deployment (QFD) or Process Failure Mode and Effects Analysis (PFMEA).

- This integration ensures a holistic approach to product quality and reliability, addressing potential issues from both the design and process perspectives.

Conducting a DFMEA is a vital part of the design process, enabling teams to proactively address potential failure points. Through a structured, methodical approach, DFMEA helps in minimizing risks associated with product failures, thereby enhancing the overall quality and reliability of the product.

Utilizing DFMEA Findings in the Design Process

Incorporating insights from Design Failure Mode and Effects Analysis (DFMEA) into the design process is crucial for enhancing product quality and reliability. These insights help refine the product design, ensuring it meets safety standards and functions effectively under various conditions.

Revising Design Based on DFMEA Insights: The insights from DFMEA guide the necessary revisions in the product design. Designers can adjust materials, modify component dimensions, or add features to improve safety and functionality.

Building Robust Designs: Understanding potential failure points allows designers to create more durable products. Such designs are essential for products used in challenging environments or those with extended lifespans.

Improving Safety and User Experience: Addressing potential failure modes directly impacts user safety. Enhancing safety features not only ensures compliance with standards but also elevates the user experience, making products more reliable and user-friendly.

Informed Material Selection: DFMEA provides valuable information on how different materials behave and contribute to potential failures. This knowledge assists designers in choosing the right materials for durability, performance, and cost-effectiveness.

Adjusting Manufacturing Processes: Insights from DFMEA can influence manufacturing decisions. Understanding how various production methods affect product reliability guides the selection of appropriate manufacturing techniques and quality control measures.

Encouraging Innovative Solutions: The process of identifying and addressing failure modes can lead to innovative design solutions. This creative problem-solving enhances the product's design and functionality.

Enhancing Team Collaboration: The DFMEA process promotes better communication among various departments. Sharing understanding of potential failures encourages teamwork and a unified approach to product development.

Assisting with Regulatory Compliance: Documenting DFMEA findings is crucial for proving compliance with industry regulations, especially in sectors like healthcare and automotive. This documentation is often critical during product certification processes.

Creating a Knowledge Base for Future Projects: Documenting the outcomes of DFMEA provides a valuable resource for future design projects. This repository of information helps avoid past mistakes and guides new product development.

Establishing a Continuous Improvement Cycle: Integrating DFMEA findings into design is part of an ongoing improvement process. This approach allows for constant refinement of the product, adapting to new customer needs and technological developments.

This integration of DFMEA into the design process reflects a commitment to superior product development and continuous enhancement. The insights guide designers in creating high-quality, reliable, and safe products that align with customer expectations and market demands.

Incorporating Test Pads in Design

The incorporation of test pads in product design, especially in electronics, is a fundamental strategy that significantly influences the overall success and sustainability of a product. These small, conductive areas on a PCB play a multifaceted role beyond basic testing and diagnostics, intricately woven into the fabric of product development.

Role of Test Pads

Optimizing Design for Testability: Integrating test pads requires designers to consider testability as a core aspect of the design process. This approach ensures that the design accommodates easy testing, leading to more efficient troubleshooting and maintenance. It necessitates a forward-thinking design strategy where testability is as important as functionality.

Reducing Time-to-Market: Test pads can substantially reduce the time taken to move a product from the design phase to the market. Efficient testing facilitated by these pads allows for rapid identification and rectification of design issues. This swift turnaround is crucial in industries where time-to-market is a key competitive factor.

Cost Management in Production: By streamlining the testing process, test pads help in reducing production costs. Automated testing enabled by test pads reduces the need for extensive manual testing, thereby cutting down on labor costs and minimizing human error.

Supporting Firmware and Software Testing: In smart devices, test pads are not only used for hardware testing but also play a crucial role in firmware and software updates and troubleshooting. They provide physical points for loading, updating, and debugging software, an essential aspect of modern electronic product maintenance.

Facilitating End-of-Line Testing: Test pads are vital in the final stages of the manufacturing process. End-of-line testing, which ensures that the product leaving the production line meets all specifications, relies heavily on test pads for efficient and accurate testing.

Enabling Field Service and Repairs: Beyond manufacturing, test pads are invaluable during field service and repair work. They provide service technicians with access points for diagnostics and repairs, thereby extending the life of the product and enhancing customer satisfaction.

Contributing to Environmental Sustainability: Efficient testing and diagnostics facilitated by test pads can contribute to the environmental sustainability of a product. By enabling easier repair and maintenance, the lifespan of products is extended, reducing waste and the need for frequent replacements.

Quality Assurance in Complex Systems: In complex electronic systems, such as those found in aerospace or medical devices, test pads are critical for ensuring that every component functions as intended. The high standards required in these industries make test pads indispensable for comprehensive quality assurance.

Customization for Specific Testing Requirements: Test pads can be customized according to the specific testing needs of a product. For instance, in high-frequency applications, the design and placement of test pads might be optimized for minimal signal interference and accurate measurement.

Incorporating test pads into design transcends basic testing functions, impacting various facets of product development, manufacturing, and maintenance. This integration guarantees the quality and reliability of electronic products while optimizing the entire lifecycle, from design to field service. By embedding efficiency and sustainability into the product's DNA, it ensures a streamlined and eco-friendly approach.

Designing with Test Pads

The design considerations for integrating test pads in electronic devices, such as smartphones, are multifaceted and require a detailed and strategic approach.

Enhanced Design Considerations for Test Pads

Environmental and Operational Conditions: The design of test pads must account for the environmental and operational conditions under which the device will operate. This includes considerations for temperature variations, humidity, and exposure to dust or chemicals, which might affect the reliability and accessibility of the test pads.

Compatibility with Various Testing Protocols: Test pads should be versatile enough to accommodate different testing protocols. This includes electrical testing and any specific firmware or software diagnostics that might be necessary during the product's lifecycle.

Integration with Design Software Tools: Utilizing advanced PCB design software tools can aid in the optimal placement and sizing of test pads. These tools can simulate different scenarios and provide insights on the best strategies for integrating test pads without compromising the overall design.

Minimizing Signal Interference: In complex circuits like those in smartphones, careful consideration must be given to minimize signal interference. The placement of test pads should be such that it does not interfere with the normal operation of the circuit, especially in high-frequency applications.

Customization for Specialized Applications: In certain specialized applications, such as medical devices or aerospace electronics, test pads may need to be customized to meet specific standards or testing requirements. This might involve unique materials, shapes, or configurations tailored to specialized testing equipment.

The smartphone industry provides a clear example of how critical the strategic integration of test pads is for the success of a product.

Adapting to Miniaturization Trends: As smartphones continue to evolve with a trend towards miniaturization, the design of test pads must also adapt. This involves creating smaller, more efficient test pads that can fit into increasingly compact spaces without compromising their functionality.

Role in Enhancing Wireless Charging: For smartphones with wireless charging capabilities, test pads play a vital role in ensuring the reliability of charging circuits. Their placement must account for the unique challenges posed by wireless power transfer systems.

Contribution to Fast-Paced Development Cycles: In the fast-paced smartphone industry, where development cycles are short, test pads facilitate quick iterations and modifications. They allow for rapid testing and validation of design changes, which is crucial for staying competitive.

Impact on Consumer Perceptions of Quality: The effective use of test pads in ensuring product reliability has a direct impact on consumer perceptions of quality. In a market where brand reputation is key, the assurance of quality that comes from thorough testing using test pads can be a significant differentiator.

Support for Software and Hardware Integration: In modern smartphones, the integration of hardware and software is critical. Test pads enable efficient testing of this integration, ensuring that both hardware components and software applications function seamlessly together.

As we have discussed, the strategic design and integration of test pads are fundamental to the functionality and reliability of complex electronic products like smartphones. Their role extends beyond basic testing, influencing various aspects of design, manufacturing, and post-sale service. The careful consideration of placement, size, durability, and environmental factors in designing with test pads is essential in creating products that meet the high standards of performance and reliability expected in today's market. In industries such as smartphone manufacturing, where the margin for error is minimal, the effective use of test pads becomes a crucial component of the design process, significantly contributing to the product's success and customer satisfaction.

By employing a comprehensive approach to designing with test pads, manufacturers can guarantee optimal functionality and long-term performance of their products. This enhances the overall user experience and reinforces the brand's reputation for quality. This attention to detail in the design phase, with a focus on testability and quality assurance, is a key factor in the longevity and success of high-tech electronic products.

Ensuring Design Maturity: Criteria for Deciding When the Design is Ready for the Next Phase

Determining the maturity of a design and its readiness to transition to the next phase of product development is a critical decision that impacts the overall success of the project. This decision hinges on a set of comprehensive criteria that assess various aspects of the design.

Completion of Design Specifications: The design should fully meet the outlined specifications. This includes compliance with all functional requirements, performance targets, and user needs. The design should be evaluated to ensure it aligns with the initial project goals and objectives.

Successful Integration of DFMEA Insights: The design should incorporate all relevant insights and modifications arising from the Design Failure Mode and Effects Analysis (DFMEA). This integration ensures that potential failure modes have been identified and mitigated, enhancing the reliability and safety of the product.

Verification and Validation Tests: The design must undergo thorough verification and validation tests. Verification ensures the design meets the specified requirements, while validation confirms that the product fulfills its intended purpose and meets user needs. Passing these tests indicates that the design is functionally and operationally sound.

Prototype Evaluation: If prototypes have been developed, they should be rigorously evaluated for performance, durability, and usability. Feedback from prototype testing is crucial in determining if the design has achieved its intended purpose and whether it is ready to proceed to mass production or the next development stage.

Compliance with Regulatory Standards: The design must comply with all relevant industry and regulatory standards. This compliance is essential for legal and safety reasons and often requires extensive documentation and certification.

Manufacturability Assessment: The design should be assessed for its manufacturability. This includes evaluating the ease of production, availability of materials, cost-effectiveness, and the feasibility of manufacturing processes. A design that is difficult or too costly to manufacture may need further refinement.

Sustainability and Environmental Impact: In contemporary product development, the design should be evaluated for its sustainability and environmental impact. This involves assessing the use of eco-friendly materials, energy efficiency, and the product's carbon footprint.

Stakeholder Approval: All key stakeholders, including design teams, management, and clients (if applicable), should review and approve the design. Stakeholder approval is essential to ensure that the design meets the broader objectives and expectations of the company and its customers.

Risk Management: Assess the design for any residual risks. Even with thorough testing and analysis, some risks may persist. The design should only move forward if these risks are within acceptable levels and have been clearly communicated to all relevant parties.

Documentation and Record-Keeping: Comprehensive documentation of the design process, testing results, and compliance certifications should be complete and up-to-date. Proper documentation is not only critical for reference and regulatory purposes but also for ensuring transparency and traceability in the design process.

Ensuring design maturity requires a multifaceted assessment that covers technical, operational, and compliance aspects. A design is ready for the next phase only when it has met all specified criteria, indicating that it is reliable, manufacturable, compliant with regulations, and aligned with market and user needs. This thorough evaluation process is crucial for minimizing risks and ensuring the success of the product as it moves forward in the development lifecycle.

Preparing for Manufacturing and Beyond

The transition from the design phase to manufacturing is a critical juncture in product development. It involves a series of steps and considerations to ensure a seamless shift, aligning the conceptual design with practical manufacturing processes.

Transitioning from Design to Manufacturing

Final Design Review and Approval: Before transitioning to manufacturing, conduct a final review of the design. This should confirm that all specifications, requirements, and standards are met. Approval from all relevant stakeholders, including design teams and management, is crucial.

Design Freeze and Locking Specifications: Establish a design freeze, which means no further changes to the design are allowed unless absolutely necessary. This step ensures stability in the design and allows manufacturing teams to proceed with confidence.

Manufacturability Analysis: Conduct a detailed manufacturability analysis. This involves evaluating the design for ease of manufacturing, identifying potential production challenges, and determining the feasibility of mass production. Modifications may be necessary to optimize the design for manufacturing efficiency.

Supplier Engagement and Material Sourcing: Engage with suppliers to ensure the availability of materials and components. Finalize sourcing agreements and confirm lead times to align with the production schedule.

Tooling and Equipment Preparation: Develop and prepare necessary tooling and equipment for production. This might involve creating molds for injection molding, setting up CNC machines, or calibrating assembly line robotics.

Quality Control Systems Setup: Establish quality control measures and systems. This includes defining inspection criteria, setting up testing procedures, and ensuring compliance with quality standards throughout the manufacturing process.

Documentation and Knowledge Transfer

Comprehensive Documentation Package: Compile a comprehensive documentation package that includes design files, specifications, DFMEA reports, test results, and compliance certificates. This package serves as a blueprint for manufacturing and ensures all teams have access to critical information.

Knowledge Transfer Sessions: Organize knowledge transfer sessions between the design and manufacturing teams. These sessions should cover critical aspects of the design, potential challenges, and any unique elements that require special attention during production.

Creating Standard Operating Procedures (SOPs): Develop SOPs for manufacturing processes based on the design. These procedures should provide clear instructions for production, assembly, and quality control.

Training for Manufacturing Teams: Provide training to manufacturing teams, especially if the product involves new technologies or complex assembly processes. This training ensures that the workforce is equipped with the necessary skills and knowledge to produce the product accurately and efficiently.

Setting Up Feedback Mechanisms: Establish feedback mechanisms between manufacturing and design teams. Continuous feedback during the initial production stages can identify any unforeseen issues and allow for timely adjustments.

Pilot Runs and Scale-Up Planning: Start with pilot production runs to test the manufacturing process and identify any adjustments needed before full-scale production. Use findings from these initial runs to plan for scale-up and mass production.

Preparing for manufacturing and beyond requires meticulous planning, detailed documentation, and effective communication between design and manufacturing teams. By carefully managing this transition, manufacturers can ensure that the product adheres to the design vision and is also produced efficiently, cost-effectively, and to the highest quality standards. This preparation is pivotal in transforming a well-designed prototype into a successful, market-ready product.

Forecasting Future Design Changes: Anticipating and Planning for Iterations Post-Manufacturing

Forecasting future design changes is an integral part of the product development process, especially in today's rapidly evolving technological landscape. Anticipating and planning for these changes ensures that the product remains competitive, relevant, and adaptable to evolving market demands and technological advancements.

Understanding Market and Technological Trends

Market Analysis: Regularly analyze market trends and consumer behaviors to predict future needs and preferences. This involves staying informed about emerging technologies, competitor products, and changing consumer expectations.

Technology Monitoring: Keep abreast of new technologies and materials that could enhance product performance or reduce manufacturing costs. This includes advancements in manufacturing processes, electronics, software, and materials science.

Incorporating Flexibility in Design

Modular Design: Implement a modular design approach where possible. This allows for easier updates and upgrades of certain components without overhauling the entire product, thereby extending its lifecycle and adaptability.

Design for Upgradability: Plan for potential future upgrades during the design phase. This may include adding extra capacity, compatibility with new standards, or provisions for additional features that could be introduced later.

Feedback Loops and Continuous Improvement

Customer Feedback Mechanisms: Establish robust channels for customer feedback. This includes leveraging social media, customer support interactions, and product reviews to gather insights on product performance and areas for improvement.

Post-Launch Product Analysis: Conduct regular assessments of the product post-launch to identify any issues or areas for enhancement. Use these findings as a basis for future design iterations.

Planning for Regulatory and Compliance Updates

Regulatory Forecasting: Stay updated with impending regulatory changes and plan for compliance in future product versions. This is particularly crucial in heavily regulated industries such as healthcare and electronics.

Compliance Flexibility: Design with regulatory flexibility in mind, particularly for international markets, where regulations may vary significantly.

Collaboration with Stakeholders

Engaging with Suppliers and Partners: Maintain strong relationships with suppliers and partners to gain insights into potential improvements in components or materials that could be incorporated into future product iterations.

Cross-Functional Team Discussions: Foster an environment of open communication and collaboration among different teams within the organization. Regular discussions between design, manufacturing, sales, and marketing teams can provide valuable perspectives on potential future design changes.

Risk Management and Scenario Planning

Risk Assessment for Future Changes: Conduct risk assessments for potential future changes. This involves evaluating the impact of these changes on current product performance, manufacturing processes, and market positioning.

Scenario Planning: Engage in scenario planning to prepare for various future possibilities. This includes creating plans for best-case and worst-case scenarios regarding technological shifts, market changes, and consumer trends.

Alpha and Beta prototypes serve as key milestones in the product development process, transforming digital designs into tangible entities. They provide invaluable insights into the real-world performance, manufacturability, and user interaction of the product. This transition from digital to tangible is a gateway to extensive testing, user feedback, and iterative improvements that shape the final product.

In the forthcoming discussion, we will delve into the nuances of Alpha and Beta prototyping, exploring how these stages bridge the gap between design forecasts and the actual product that will eventually reach the market.

Alpha and Beta Prototypes – From Digital to Tangible

In the captivating journey of product development, prototyping stages stand as critical milestones, transforming abstract ideas into tangible realities. The journey through Alpha and Beta prototypes is similar to a sculptor chiseling a formless block into a defined shape, where each stroke brings clarity and definition to the final product.

Alpha Prototypes: The First Glimpse into Reality

The Alpha prototype represents the first substantial step in bringing a digital design to life. It is a preliminary version of the product, often rough and unrefined, but it serves a crucial purpose. This stage is about testing the feasibility of the concept, ensuring that the basic functions and mechanics of the product are viable.

Functionality Over Form: At this stage, the primary focus is on functionality rather than aesthetics. The Alpha prototype may not resemble the final product in appearance, but it plays a vital role in validating the core design principles and functionalities.

Iterative Process: Alpha prototyping is inherently iterative. Each version seeks to improve upon its predecessor, gradually refining the design based on testing results and feedback.

Consider Alpha as the ultimate product, encompassing all design considerations. While it may require additional time, its significance cannot be overstated. Always aim to design and deliver the Alpha prototype flawlessly on the first attempt. Thus, Alpha should align with form, fit, and functionality. This aligns with the concept of Minimum Viable Product (MVP), which is widely embraced by funding agencies. Consequently, Alpha should encompass all aspects such as design for manufacturability, Design to Cost, Design to Compliance, and Design to Reliability.

Beta Prototypes: Refinement and User Interaction

Once the Alpha prototype has proven the concept's feasibility, the process advances to the Beta stage. Beta prototypes are more advanced, closely resembling the final product in both form and function. This stage is pivotal in refining user experience and identifying any remaining issues before mass production.

Polished and Functional: Beta prototypes are more polished and are expected to function as the final product would. This stage involves comprehensive testing, including user testing, to refine the design and functionality.

Pre-Market Validation: This prototype is also used for pre-market testing, providing valuable insights into how potential users interact with and respond to the product.

Consider the development of a smart thermostat as an illustrative example. In its Alpha stage, the focus is on creating a functional unit that can regulate home temperature based on user inputs and sensor data. The initial Alpha prototype may be a basic assembly of circuit boards and sensors, housed in a rudimentary case. It doesn't need to look elegant; it just needs to demonstrate that the core idea – a thermostat that can be controlled remotely and learn user preferences – is feasible.

During this phase, engineers and designers test various components, such as temperature sensors, connectivity modules for Wi-Fi, and algorithms for learning user behavior. They likely encounter challenges like ensuring accurate temperature readings and seamless integration of the software with the hardware.

Transitioning to the Beta prototype, the smart thermostat starts to take on a more refined form, resembling the sleek, user-friendly device intended for consumers. This prototype is more sophisticated, featuring a refined user interface on a polished touchscreen, a more elegant casing, and enhanced software.

In the Beta phase, the prototype is likely placed in real-world settings, like homes or offices, to gather feedback on its usability, aesthetics, and functionality. User interaction plays a crucial role at this stage, helping the team understand how people interact with the device, how intuitive the controls are, and how effectively it integrates into daily routines.

This iterative process from Alpha to Beta prototypes is crucial in resolving issues such as improving the sensitivity of the thermostat, enhancing the user interface for a more intuitive experience, and ensuring reliable connectivity for remote control. The smart thermostat's journey from a basic functional model to a refined, market-ready product exemplifies the essential roles that Alpha and Beta prototypes play in the product development cycle.

Transition from Design to Prototype: How Prototypes Bring Designs to Life

The transformation from a conceptual design to a physical prototype is a remarkable journey in product development. It's a phase where ideas and digital blueprints evolve into tangible, interactive models. This transition is crucial, as it moves a product from theoretical constructs into real-world applications.

Materializing Concepts: The first step in this transition is the materialization of design concepts. What were once sketches and digital renderings become physical objects. This materialization is more than just a change in form; it represents a critical leap in the development process, where designs are tested against the laws of physics, user interaction, and manufacturing constraints.

Functional Testing: Prototypes serve as a testing ground for the product's functionality. They allow designers and engineers to assess how well the product performs its intended function, identify design flaws, and make necessary adjustments. This is especially important for complex products where theoretical functionality needs thorough validation.

Ergonomics and User Experience: In transitioning to prototypes, special attention is paid to ergonomics and user experience. The prototype allows for an assessment of how the product feels and operates in a user's hands, leading to improvements in design that enhance usability and comfort.

Objectives of Each Stage: Goals and Expectations for Alpha and Beta Prototypes

Alpha Prototype Objectives

Feasibility Testing: The primary goal of the Alpha prototype is to test the feasibility of the design. It answers critical questions about whether the concept can be transformed into a viable product.

Functionality Verification: Alpha prototypes focus on verifying the core functionality of the product. They are used to ensure that the basic operations work as intended.

Identification of Design Flaws: These prototypes are essential for identifying and addressing major design flaws early in the development process. This early detection saves time and resources in later stages.

Beta Prototype Objectives

Refinement for User Interaction: Beta prototypes aim to refine the product for user interaction. They represent a version of the product that users can interact with and provide feedback on, which is crucial for the next round of refinements.

Testing in Real-World Scenarios: Unlike Alpha prototypes, Beta prototypes are tested in conditions that closely mimic the real world. This stage assesses how the product performs under typical usage scenarios, revealing issues that might not have been apparent in the controlled environment of the Alpha stage.

Market Readiness Assessment: Beta prototypes are instrumental in evaluating the product's readiness for the market. This includes evaluating aesthetic appeal, user interface, durability, and overall user experience. It's a critical phase for making final adjustments before committing to mass production.

Gathering Detailed User Feedback: An essential objective of Beta prototyping is to gather detailed feedback from potential users or stakeholders. This feedback is invaluable in understanding user expectations, preferences, and potential issues from a user's perspective.

Compliance and Safety Testing: Beta prototypes are also used for compliance testing with relevant industry standards and safety regulations. Ensuring that the product meets these requirements is vital before it can be released to the market.

Significance of Prototypes in Product Development

Prototyping is the bridge between imagination and reality. The transition from design to prototype is where abstract ideas are confronted with practical constraints, and where visionary concepts are tested against user needs and manufacturing realities.

The Alpha and Beta stages of prototyping are essential for different reasons. The Alpha stage is about turning ideas into something tangible and functional, a process that is often filled with discoveries and challenges. On the other hand, the Beta stage is about refinement, user interaction, and preparing the product for the real world. It's where the product is polished, user feedback is integrated, and the final touches are added to ensure market success.

Navigating through the world of prototyping presents a unique set of challenges. Each step of the journey from conceptual design to a tangible prototype is fraught with potential obstacles that require careful consideration and strategic problem-solving. Understanding these common challenges and how to address them is essential for a smooth prototyping process, ensuring the transition from an idea to a functional prototype is as seamless as possible.

Overcoming Challenges in Prototyping

Material Selection and Sourcing Issues: One of the first hurdles in prototyping is selecting and sourcing the right materials. Materials must not only be suitable for the intended functionality of the prototype but also be readily available and cost-effective. Addressing this involves thorough research into material properties and building relationships with reliable suppliers.

Balancing Design Complexity with Manufacturability: Prototypes often push the boundaries of design, leading to complexity that can be challenging to manufacture. Overcoming this requires a balance between innovation and practical manufacturability, often necessitating design compromises or the exploration of alternative manufacturing methods.

Time Constraints and Rapid Prototyping Needs: Time is a critical factor in prototyping, especially in competitive markets. Rapid prototyping technologies like 3D printing can expedite the process, but managing expectations and ensuring quality within these tight timeframes remains a challenge.

Accuracy and Fidelity to Final Design: Ensuring that the prototype accurately represents the final product is crucial. This challenge involves meticulous attention to detail in the design phase and a clear understanding of the limitations and capabilities of the prototyping methods being used.

Cost Management: Budget constraints are a common issue in prototyping. Effective cost management involves strategic planning, considering cost-effective prototyping methods, and iterative testing to prevent costly redesigns in later stages.

Integration of Different Components: For prototypes involving multiple components or systems, ensuring seamless integration can be challenging. This requires a holistic approach to design and testing, ensuring compatibility and functionality across all components.

User-Centered Design and Feedback Incorporation: Prototyping must often take into account user feedback and ergonomic considerations. The challenge lies in incorporating this feedback into the prototype design without losing sight of the core functionalities and technical requirements.

Prototype Testing and Validation: Comprehensive testing of prototypes to validate their functionality, durability, and performance can be complex. Developing effective testing protocols and adapting them as the prototype evolves is essential for accurate validation.

Environmental and Regulatory Compliance: For certain products, meeting environmental standards and regulatory compliance during the prototyping phase can be challenging. This requires staying updated with relevant regulations and integrating compliance into the design and testing process.

Scalability to Full-Scale Production: While this stage is not about full-scale production, it's important to consider how the prototype will eventually scale up. Challenges arise in ensuring that the prototype is designed in a way that is feasible for mass production, both in terms of design and material use.

Overcoming challenges in prototyping demands a blend of creativity, technical knowledge, strategic planning, and adaptability. Each obstacle presents an opportunity to refine the prototype, ensuring it is a robust representation of the final product. By proactively addressing these common challenges, developers can streamline the prototyping process, paving the way for a successful transition from a visionary concept to a market-ready product.

Cost Management:

Managing costs in prototype development is about making informed choices that balance the interplay between cost, time, and quality. It requires a strategic approach, careful resource allocation, and a willingness to adapt and reevaluate as the project evolves. Ultimately, successful cost management in prototyping is not just about cutting expenses but making smart investments that enhance the value and feasibility of the final product.

It's about foresight, flexibility, and a deep understanding of the prototyping process, ensuring that each step taken is a stride towards innovation without compromising the integrity of the design or the practical aspects of production. This delicate balance is what transforms good ideas into viable, market-ready products within the confines of budgetary realities.

Understanding the Cost-Quality-Time Triangle: The core of cost management in prototyping lies in navigating the cost-quality-time triangle. These three elements are interconnected: reducing costs may impact quality or extend timelines, while accelerating development time can increase costs or compromise quality.

The key is to find an optimal balance that aligns with the project's goals and resources.

Strategic Planning and Budget Allocation: Effective cost management begins with strategic planning. This involves defining clear objectives, setting realistic budgets, and allocating resources judiciously. It's crucial to prioritize aspects of the prototype that are critical to its functionality and user experience and allocate funds accordingly.

Leveraging Rapid Prototyping Techniques: Utilizing rapid prototyping methods such as 3D printing or CNC machining can be a cost-effective way to create prototypes. These technologies not only reduce the time required to produce a prototype but also often lower the costs associated with traditional manufacturing methods like molding or casting.

Iterative Design and Prototyping: An iterative approach to design and prototyping can significantly manage costs. By creating simpler, less expensive prototypes in the initial stages and gradually adding complexity, it's possible to identify and resolve issues early on, preventing costly redesigns later in the process.

Material Selection and Sourcing Efficiency: Choosing the right materials for a prototype is a balancing act between cost, performance, and availability. Sometimes, using less expensive or more readily available materials in the early stages of prototyping can save costs without majorly impacting the overall functionality.

Outsourcing vs. In-House Prototyping: Deciding between outsourcing prototyping and keeping it in-house is another crucial aspect of cost management. Outsourcing can reduce initial capital expenditures but may increase unit costs. Conversely, in-house prototyping requires upfront investment in equipment and expertise but can be more cost-effective in the long run, especially for ongoing projects.

Minimizing Design Revisions through Effective Testing: Comprehensive and effective testing of prototypes can help minimize costly design revisions. By thoroughly testing the prototype under real-world conditions and gathering user feedback, necessary adjustments can be identified and implemented efficiently.

Scalability Considerations: While focusing on prototyping, it's also important to keep an eye on the scalability of the design to full-scale production. Designs that are cost-effective to prototype but expensive to manufacture at scale can lead to challenges in later stages.

Transparent Communication with Stakeholders: Keeping open lines of communication with all stakeholders involved in the prototyping process is vital. This transparency helps in managing expectations, making informed decisions, and aligning the project goals with the available budget.

Ensuring Timelines

Ensuring the prototyping phase stays on schedule is critical in product development. The timely completion of prototypes involves managing resources efficiently, maintaining project momentum, and capitalizing on market opportunities. A combination of meticulous planning, agile management, and proactive problem-solving is essential for this phase to progress smoothly.

Effective Strategies for Maintaining Timelines in Prototyping

Detailed Planning with Clear Timelines: Start with a comprehensive plan that outlines every stage from design conception to final prototype testing. Include estimated durations and key milestones to create a roadmap for the project, helping to identify potential delays early on.

Setting Achievable Deadlines: Balance ambition with realism when setting deadlines. They should challenge the team but remain attainable, based on an accurate assessment of the tasks and resources involved.

Consistent Progress Reviews and Adjustments: Conduct regular reviews to monitor progress. These sessions help track milestones, assess the work completed, and identify deviations from the plan, allowing for timely adjustments.

Streamlined Communication and Coordination: Maintain clear and continuous communication among team members and stakeholders. Use regular updates, meetings, and collaborative tools for alignment on the project's status and schedule changes.

Proactive Risk Management: Anticipate potential risks and develop contingency plans. This approach involves predicting issues like technical challenges or supply chain delays and planning effective responses.

Prioritizing Tasks and Efficient Resource Allocation: Focus on prioritizing tasks that have a significant impact on the project timeline. Allocate resources, including personnel and equipment, to critical activities that directly influence the timeline.

Adopting Rapid Prototyping Methods: Rapid prototyping techniques such as 3D printing can expedite the development process, enabling quicker iterations and faster testing.

Strategic Outsourcing: When appropriate, outsource specific prototyping tasks. Choosing expert external partners for specialized tasks can speed up those aspects of the project.

Embracing Flexibility and Adaptability: Stay flexible and adapt to challenges. This may mean reorganizing tasks, reallocating resources, or adjusting prototype features to overcome hurdles efficiently.

Incorporating User Feedback Effectively: Manage the integration of user or stakeholder feedback in a way that gleans valuable insights without significantly delaying the process. Structured testing and focused feedback sessions can be beneficial.

Balancing Workload and Team Well-being: Ensure the team's workload is manageable to maintain productivity. Regularly assess the team's well-being to prevent burnout and errors, which can lead to delays.

Utilizing Project Management Tools: Implement project management software to enhance task tracking and team collaboration. These tools offer a central platform for progress monitoring and communication, aiding in keeping the project on target.

Understanding and overcoming challenges in prototyping sets the stage for more advanced and efficient methods of realization, like 3D Printing. This transition from traditional approaches to innovative technologies marks a significant step in the evolution of prototyping.

3D Printing in Prototyping

3D printing stands at the forefront of this evolution, revolutionizing how prototypes are developed. This technology shifts the paradigm from conventional, often labor-intensive processes to a more agile and versatile approach. By constructing objects layer by layer, 3D printing allows for the materialization of intricate designs that were once deemed too complex or unattainable with traditional manufacturing. This capability not only accelerates the prototyping process but also opens up new avenues for creativity and innovation, enabling designers and engineers to push the boundaries of what can be physically created.

Basics of 3D Printing

Layer-by-Layer Construction: 3D printing works by building objects layer by layer, using a variety of materials such as plastics, resins, or metals. This process starts with a digital 3D model, which is sliced into thin, horizontal cross-sections. The printer then creates each layer successively to form the final object.

Types of 3D Printing Technologies: There are several types of 3D printing technologies, each with its unique method of layer creation. These include Fused Deposition Modeling (FDM), Stereolithography (SLA), and Selective Laser Sintering (SLS), among others. Each technology has its strengths and is chosen based on the requirements of the prototype in terms of material, resolution, and durability.

Design Flexibility and Complexity: One of the most significant advantages of 3D printing is the design flexibility it offers. Designers are not constrained by the limitations of traditional manufacturing processes, allowing for the creation of complex and intricate designs.

Applications in Prototyping

Rapid Prototyping: 3D printing has become synonymous with rapid prototyping due to its speed and efficiency. It allows designers and engineers to quickly turn concepts into physical models, facilitating faster iterations and design refinements.

Functional Prototypes: Beyond mere form, 3D printing can produce functional prototypes that closely mimic the properties and functions of the final product, allowing for practical testing and evaluation.

Customization and Personalization: This technology is ideal for creating customized or personalized prototypes, which is particularly beneficial in industries like medical devices, where patient-specific models can be produced.

A compelling example of 3D printing in prototyping can be observed in the aerospace industry. Companies like SpaceX and Boeing utilize 3D printing for creating intricate prototypes and for manufacturing parts used in actual spacecraft and aircraft.

Prototyping Complex Parts: In aerospace, the design of certain components can be exceptionally complex, involving intricate geometries and lightweight yet strong materials. 3D printing allows for the creation of these complex prototypes quickly and accurately.

Testing and Validation: Aerospace companies frequently use 3D printed prototypes for testing and validating the design of various components. These prototypes undergo rigorous testing under conditions that simulate actual flight, including stress tests and aerodynamic assessments.

Reducing Costs and Lead Times: Traditional manufacturing methods for aerospace components can be costly and time-consuming. 3D printing significantly reduces both the cost and lead time for producing prototypes. This is especially beneficial for custom or one-off parts, where the expense of creating traditional molds or tooling is not viable.

Material Innovation: The aerospace industry often requires materials that are lightweight yet strong. 3D printing has advanced to use specialized materials, such as high-performance polymers and metal alloys, that meet these stringent requirements. Prototypes made with these materials provide valuable insights into the performance and feasibility of the design.

SpaceX's Use of 3D Printing: SpaceX, for instance, has utilized 3D printing for producing parts of its rocket engines. The ability to prototype these parts rapidly accelerates the development cycle, allowing for more frequent launches and innovations in spacecraft design. This approach has been pivotal in their quest to reduce the cost of space travel.

3D printing in prototyping is a game-changer across various industries, with the aerospace sector being a prime example. It empowers designers and engineers to push the boundaries of innovation, creating prototypes that are intricate, functional, and closer to the final product than ever before. The technology's ability to shorten development times, reduce costs, and enable material and design innovations aligns perfectly with the fast-paced, high-stakes nature of product development in the modern era. As 3D printing technology continues to evolve, its role in prototyping will undoubtedly expand, further transforming the landscape of product design and manufacturing.

Assessing When to Use 3D Printing in the Prototyping Process

3D printing, a cornerstone in modern prototyping, offers a suite of advantages that make it an invaluable tool in product development. However, like any technology, it also has its limitations. Understanding when and how to leverage 3D printing is critical for optimizing its benefits in the prototyping process.

Advantages of 3D Printing in Prototyping

Speed and Efficiency: One of the most significant benefits of 3D printing is its ability to accelerate the prototyping process. Traditional prototyping methods can take weeks or even months, whereas 3D printing can produce a prototype in a matter of hours or days. This rapid turnaround time is crucial in fast-paced industries where time to market is critical.

Complex Geometries and Customization: 3D printing excels in creating complex geometries that are challenging or impossible to achieve with traditional manufacturing methods. It allows for intricate designs, including internal structures and complex shapes, without additional cost.

Cost-Effectiveness for Small Batches: For small batch production or one-off prototypes, 3D printing is highly cost-effective. It eliminates the need for expensive tooling, molds, or setups that are typically required in traditional manufacturing.

Iterative Design: The technology is particularly well-suited for iterative design processes. Designers can quickly produce a prototype, test it, modify the design, and reprint, facilitating a rapid cycle of improvement and refinement.

Material Diversity: Recent advancements in 3D printing technology have expanded the range of materials available, including various plastics, resins, metals, and composites. This diversity allows for a wide range of mechanical and aesthetic properties in prototypes.

Limitations of 3D Printing in Prototyping

Material Properties: While the range of 3D printing materials is broad, they may not always match the properties of materials used in final production. This discrepancy can be a limitation when prototypes need to replicate the exact material characteristics of the end product.

Surface Finish and Resolution: The surface finish of 3D printed parts may require post-processing for a smooth appearance, especially when the prototype is intended for aesthetic evaluation. Additionally, the resolution of 3D printers, while constantly improving, may not always achieve the fine detail required for some applications.

Size Constraints: Most 3D printers have size limitations, restricting the dimensions of prototypes that can be printed. Larger prototypes may need to be printed in sections and assembled, which can affect structural integrity and appearance.

Cost-Effectiveness at Scale: While 3D printing is cost-effective for small runs, it may not be the most economical choice for large-scale production due to slower production speeds and higher material costs compared to traditional manufacturing methods.

Strength and Durability: Prototypes made through 3D printing may not have the same strength and durability as those made through traditional manufacturing, especially for parts that will undergo rigorous testing or functional use.

Assessing the Use of 3D Printing in Prototyping

When deciding whether to use 3D printing in the prototyping process, it's essential to consider the specific requirements of the prototype. For complex, highly customized prototypes needed in a short timeframe or where iterative design is crucial, 3D printing is an optimal choice. It's also suitable when the prototype needs to demonstrate intricate internal structures or when traditional manufacturing methods are prohibitively expensive due to low volumes.

However, for prototypes that must mimic the exact material properties of the final product, or when surface finish and structural integrity are of paramount importance, alternative methods or a combination of techniques may be more appropriate. In scenarios where large-scale production is anticipated, it's beneficial to consider prototyping methods that align more closely with the eventual manufacturing process to ensure scalability and cost-effectiveness.

Additive Manufacturing: Beyond 3D Printing

Transitioning from the realm of traditional 3D printing, Additive Manufacturing (AM) represents a significant leap in technological capabilities. Additive Manufacturing (AM), often used interchangeably with 3D printing, actually encompasses a broader range of technologies and processes that go beyond conventional 3D printing capabilities. While traditional 3D printing is a form of additive manufacturing, the term AM often implies more advanced, industrial-scale processes that offer greater complexity and precision.

Understanding Additive Manufacturing

Advanced Techniques and Materials: AM technologies, including Selective Laser Sintering (SLS), Electron Beam Melting (EBM), and Direct Metal Laser Sintering (DMLS), utilize a range of advanced materials. High-grade metals suitable for aerospace components, durable ceramics for medical devices, and high-performance composites for automotive parts are just a few examples. These materials extend far beyond the polymers and resins typically associated with basic 3D printing.

Precision and Complexity: The capabilities of AM to produce parts with intricate internal structures, complex geometries, and precise tolerances are unparalleled. This precision is particularly vital in applications where even the smallest deviation can significantly impact performance, such as in aerospace engineering or medical device manufacturing.

Industrial Scale and Efficiency: Additive Manufacturing is not confined to the realm of prototyping and small-scale production. Its adoption for industrial-scale production is growing, thanks to its efficiency and the exceptional quality of the output. Large-scale AM setups are capable of producing significant volumes of parts while maintaining the precision and complexity that AM is known for.

Layering Techniques for Enhanced Strength: Unlike traditional 3D printing, some AM processes involve specialized layering techniques that enhance the strength and durability of the final product. These techniques, such as interlacing layers at specific angles or using variable layer thickness, contribute to the mechanical properties of the printed objects, making them suitable for more demanding applications.

Integration with Digital Technologies: AM is at the forefront of the digital manufacturing revolution. It seamlessly integrates with digital design tools like CAD (Computer-Aided Design) and CAE

(Computer-Aided Engineering), enabling a more streamlined transition from digital models to physical prototypes. This integration is crucial in industries like automotive and aerospace, where complex designs originate as digital models.

Customization at Scale: One of the standout features of AM is its ability to customize products at scale. Unlike traditional manufacturing, which requires significant retooling for each custom variant, AM allows for customization without additional setup costs. This capability is especially beneficial in industries like healthcare, where patient-specific implants and prosthetics can be manufactured efficiently.

Sustainable Manufacturing Practices: AM contributes to sustainable manufacturing practices by reducing waste. Unlike subtractive methods that remove material to create a shape, AM builds structures layer by layer, using only the material necessary for the part. This efficiency not only saves material but also aligns with the growing need for environmentally conscious manufacturing processes.

Additive Manufacturing (AM) has revolutionized the approach to prototype development, offering unique capabilities that cater to the needs of various industries. The versatility of AM in prototype development can be understood in greater depth with the following insights:

Biocompatible Material Use in Medical Prototypes: In the medical field, AM's ability to utilize biocompatible materials is groundbreaking. This feature is crucial for creating prototypes of medical devices that come into direct contact with human tissue. For instance, surgical guides and organ models made from biocompatible materials can be used for pre-surgical planning, significantly improving surgical outcomes.

Aerospace Prototype Testing under Extreme Conditions: In aerospace, AM prototypes are not only used for their complex designs but also for their ability to withstand extreme conditions. Materials used in AM can be tested for high-temperature resistance, stress endurance, and aerodynamic properties, mirroring the challenging conditions encountered in aerospace environments.

Rapid Tooling for Manufacturing Prototypes: AM extends its utility to rapid tooling, which is creating tools through additive manufacturing. This approach is particularly useful in prototyping complex machinery or equipment parts, allowing for quicker transitions from design to functional testing.

Enhanced Aesthetic Possibilities in Consumer Products: For consumer products, AM brings enhanced aesthetic possibilities, allowing designers to experiment with forms and structures that would be challenging with traditional manufacturing. This capability enables the creation of visually striking prototypes that closely resemble the final product, crucial for market testing and consumer feedback.

Prototyping for Wearables and IoT Devices: In the rapidly growing sectors of wearables and IoT (Internet of Things), AM plays a significant role in prototyping small, intricate devices with embedded electronics. The precision of AM allows for the development of compact, integrated prototypes that are essential for testing the functionality of these advanced gadgets.

Sustainability in Material Usage: The eco-friendly aspect of AM extends to the use of recycled materials in prototype development. Some AM processes can utilize recycled plastics or metals, contributing to a circular economy and reducing the environmental footprint of prototyping.

On-Demand Prototyping and Inventory Reduction: AM's on-demand manufacturing capability is vital for industries requiring quick responses to market changes. This approach reduces the need for large prototype inventories, as items can be printed as needed, leading to cost savings and reduced storage requirements.

Prototyping for Custom Architectural Models: In architecture, AM allows for the creation of detailed and scale-accurate models of buildings and structures. These prototypes are essential for visualizing architectural designs and for presenting tangible models to clients and stakeholders.

Combining AM with Traditional Manufacturing: For some prototypes, AM is used in conjunction with traditional manufacturing methods. This hybrid approach can optimize the strengths of both methods, such as using AM for complex parts and traditional methods for larger, less intricate components.

Additive Manufacturing significantly broadens the scope of what is possible in prototype development. Its applications span across various industries, bringing advancements in material science, design freedom, rapid testing, and eco-friendly manufacturing processes.

Examples of Successful Additive Manufacturing in Prototyping

Additive Manufacturing (AM) has made a profound impact across various industries by revolutionizing the prototyping process. To truly appreciate the transformative power of AM, let's delve into a specific and compelling case study.

Aerospace Industry – GE Aviation's LEAP Engine Fuel Nozzle

One of the most hailed examples of AM's capabilities in prototyping and beyond is seen in the aerospace sector with GE Aviation's LEAP engine fuel nozzle.

Background: The LEAP engine, used in commercial aircraft, required a fuel nozzle that was lighter, more durable, and more efficient than traditional models. GE Aviation turned to AM for a solution.

Design Challenges: The fuel nozzle needed to withstand high temperatures and pressures while maintaining precise fuel delivery. Traditional manufacturing methods would have required the assembly of about 20 different parts, posing challenges in terms of weight, complexity, and potential points of failure.

AM Solution: GE Aviation utilized Direct Metal Laser Sintering (DMLS), an additive manufacturing technique, to create the fuel nozzle as a single piece. This approach not only eliminated the need for assembly but also allowed for a design that was 25% lighter and five times more durable than its predecessor.

Results: The AM-produced fuel nozzle was a significant success. It demonstrated superior performance, with a design impossible to achieve through traditional methods. The nozzle's single-piece construction reduced the risk of leaks or failures at assembly joints. Its innovative internal structures, made possible by AM, enhanced fuel efficiency and reduced emissions.

Impact on the Industry: This application of AM in creating a critical component for the LEAP engine marked a turning point in the aerospace industry's approach to component design and manufacturing. It showcased AM's potential not just for prototyping but for full-scale production of complex, high-performance parts.

Further Developments: Encouraged by this success, GE Aviation has continued to explore additive manufacturing for other components, signaling a broader industry shift towards AM in both prototyping and manufacturing.

This case study of GE Aviation's LEAP engine fuel nozzle exemplifies the revolutionary impact of additive manufacturing in prototyping and beyond. It highlights AM's ability to overcome complex design challenges, improve product performance, and pave the way for innovations that redefine industry standards.

Vacuum Casting for Prototypes

Transitioning from the realm of additive manufacturing, where intricate designs and complex structures come to life layer by layer, we venture into another pivotal prototyping technique: vacuum casting. This process stands out for its ability to produce high-quality, detailed replicas from original models, offering a different set of advantages in the prototyping world.

Vacuum casting, also known as urethane casting, is a versatile and efficient method for producing high-fidelity prototypes and small batches of functional parts. Its process can be broken down into several key stages:

Master Model Creation: The process begins with the creation of a master model, typically produced via 3D printing or CNC machining. This model is the exact representation of the final part and is used to create the mold. The accuracy and quality of the master model are crucial, as they directly impact the fidelity of the final cast parts.

Mold Making: Once the master model is ready, it is encased in a two-part silicone mold within a casting box. The silicone is poured around the model, then left to cure. Once cured, the mold is cut open and the master model is removed, leaving a hollow cavity in the shape of the part.

Vacuum Casting: The actual casting process begins by placing the silicone mold in a vacuum chamber. A chosen polyurethane resin is mixed, de-gassed, and then poured into the mold. The vacuum environment is critical as it ensures that the resin fills the entire mold cavity without any air bubbles, which could compromise the quality of the final part.

Curing and Part Removal: After the resin is poured, the mold is placed in a heated chamber to cure. Once the resin has hardened, the mold is opened and the cast part is removed. The silicone mold can be used multiple times, allowing for multiple copies of the part to be produced.

Finishing Touches: The final stage involves cleaning and finishing the cast parts. This may include sanding, painting, or other surface treatments to achieve the desired aesthetic and functional properties.

Advantages of Vacuum Casting

High-Quality Replicas: Vacuum casting is renowned for its ability to produce parts with high levels of detail and excellent surface finish. The use of silicone molds allows for the replication of fine features and textures from the original model.

Material Versatility: A wide range of polyurethane resins are available for vacuum casting, mimicking various material properties from rigid and tough to soft and flexible. This versatility makes it suitable for a variety of applications.

Cost-Effectiveness for Small Batches: For small batch production, vacuum casting is a cost-effective alternative to injection molding, as it does not require expensive tooling. It's particularly suitable for low-volume production or for bridging the gap between prototyping and full-scale production.

Speed and Efficiency: The process is relatively quick, with short lead times for producing parts. This speed is beneficial in scenarios where time-to-market is a critical factor.

Vacuum casting is a valuable technique in the prototyping toolkit, particularly suited for small-scale production and when high-quality, detailed replicas are needed. Its ability to quickly and efficiently produce parts with a range of material properties and surface finishes makes it a preferred choice for many industries.

Assessing the suitability of vacuum casting for different types of prototypes requires a nuanced understanding of this method's capabilities and limitations. This evaluation is crucial in determining whether vacuum casting is the most appropriate technique for a given prototyping project.

Suitability of Vacuum Casting Based on Prototype Requirements

Complexity and Detail: Vacuum casting is particularly well-suited for prototypes that require a high level of detail and surface finish. Its ability to capture intricate textures and features from the master model makes it ideal for prototypes where aesthetic detail and fidelity to the original design are paramount.

Material Properties: The range of polyurethane resins available for vacuum casting allows for the simulation of various material properties, from rubber-like flexibility to rigid plastics. Prototypes that need to mimic the look and feel of specific materials, or that require certain functional properties like translucency, can benefit greatly from vacuum casting.

Batch Size and Production Volume: Vacuum casting is most cost-effective for small to medium-sized production runs. It offers an economical solution for producing a limited number of parts, especially when compared to the higher setup costs associated with injection molding. For projects requiring a small batch of high-quality prototypes, vacuum casting is an excellent choice.

Speed and Turnaround Time: Projects that require a quick turnaround can benefit from vacuum casting. Since the process doesn't require complex or expensive tooling, the time from design to finished prototype is relatively short, making it suitable for projects with tight deadlines.

Prototype Functionality: While vacuum casting produces parts that closely mimic final products in look and feel, it's important to consider the functional requirements of the prototype. For prototypes that are subjected to extensive functional testing or that need to withstand high-stress environments, the mechanical properties of vacuum cast parts might not always be sufficient.

Prototype Lifecycle: Vacuum casting is appropriate for prototypes that are used in the short to medium term. The silicone molds used in vacuum casting have a limited lifespan and can deteriorate over time or after a certain number of parts have been produced. For long-term use or very high production volumes, other methods might be more suitable.

Cost Considerations: Budget constraints play a significant role in determining the suitability of vacuum casting. For projects where budget is a limiting factor, vacuum casting offers a balance between cost and quality, especially for prototypes that require high-quality finishes and materials similar to the final product.

Environmental Considerations: For organizations with a focus on sustainability, vacuum casting can be a more environmentally friendly option compared to some traditional manufacturing methods. The process produces less waste, and the materials used are often less environmentally impactful.

Vacuum casting, with its versatile applications, has found a significant place in various industries for product development. Its unique process and advantages have made it an indispensable tool in scenarios where precision, detail, and quality are crucial. Exploring real-world applications of vacuum casting will illuminate its practical utility and effectiveness.

Consumer Electronics and Gadgets

In the consumer electronics industry, vacuum casting is often employed to create high-fidelity prototypes of gadgets such as smartphones, wearables, and remote controls. These prototypes need to replicate the final product's aesthetic and tactile feel closely. Vacuum casting allows designers to experiment with different textures and finishes, creating prototypes that look and feel like the final product. This capability is essential for user testing and marketing purposes, where the first impression of a product can be pivotal.

Automotive Industry

The automotive sector frequently utilizes vacuum casting for the development of prototypes for interior components such as knobs, buttons, and dashboards. Vacuum casting enables the production of parts that mimic the appearance and feel of various materials like soft-touch plastics and faux leather, which are commonly used in vehicle interiors. This process is particularly useful for bespoke or limited-edition models, where creating traditional tooling for a small number of parts would be economically unfeasible.

Medical Device Prototyping

In medical device prototyping, the precision and safety of prototypes are paramount. Vacuum casting is used to create highly accurate models of medical devices, such as surgical tools and equipment housings. The ability to use biocompatible materials in the casting process is a significant advantage, allowing for the creation of prototypes that can be used in clinical settings for trials or training purposes.

Custom Architectural Models

Architects and designers use vacuum casting to create detailed scale models of buildings and structures. These models are crucial for client presentations and exhibitions. Vacuum casting provides a high level of detail and allows for the incorporation of different textures and materials, enhancing the model's realism and impact.

Vacuum casting is not merely limited to prototyping; it's also used for rapid tooling applications. In short-run production, especially for customized products or limited editions, vacuum casting serves as an efficient method to produce small batches of parts without the need for expensive and time-consuming tooling.

These real-world applications showcase vacuum casting's versatility and effectiveness across various sectors. Whether it's creating lifelike prototypes for consumer electronics, detailed components in the automotive industry, precise medical device models, realistic props for the entertainment industry, architectural models, or short-run production tooling, vacuum casting proves to be a valuable asset in product development. Its ability to replicate complex designs with high fidelity and its adaptability to a range of materials make it an ideal choice for projects where detail, precision, and quality are of utmost importance.

PCB Fundamentals: Introduction to Printed Circuit Board Design and Fabrication

Shifting focus from the broad applications of vacuum casting in various industries, we now delve into the intricacies of Printed Circuit Board (PCB) building, a fundamental aspect of electronic prototyping. PCBs form the backbone of modern electronics, and understanding their design and fabrication is crucial for anyone involved in electronic product development.

Essence of PCBs in Electronics: At its core, a Printed Circuit Board is the platform upon which electronic components are mounted and interconnected. PCBs are used in virtually all electronic products, from simple gadgets like digital watches to complex systems like computer motherboards and telecommunications equipment.

Design Process: The design of a PCB starts with a schematic diagram, which is a blueprint of the circuit. This diagram includes all the components and their connections. Using PCB design software, this schematic is translated into a layout that dictates the physical placement of components and the routing of electrical traces on the board.

Material Selection: The most commonly used material for PCBs is FR4, a type of fiberglass-reinforced epoxy laminate. FR4 provides a balance of mechanical sturdiness, electrical insulation, and affordability. However, other materials like metal-core boards for high-power applications or flexible PCBs for specific use-cases are also utilized.

Layering and Stackup: PCBs can be single-sided (one copper layer), double-sided (two copper layers), or multi-layered. Multi-layered PCBs, which have multiple layers of copper separated by insulating material, are used for complex, high-density circuits. The design of the layer stackup is a crucial decision in the PCB design process, impacting the board's electrical performance and manufacturability.

Fabrication Process: The fabrication of a PCB involves several steps, including applying a copper layer to the substrate, circuit patterning (often done through photolithography), etching away excess copper, drilling holes for component leads, and applying solder mask and silkscreen layers for component labels and protection.

Surface Mount Technology (SMT) and Through-Hole Technology (THT): These are two primary methods for mounting components onto PCBs. SMT involves directly mounting components onto the surface of the board, allowing for more compact and dense designs. THT involves inserting component leads into pre-drilled holes on the board and is often used for larger components or parts that endure mechanical stress.

Testing and Quality Assurance: After fabrication, PCBs undergo rigorous testing to ensure they meet the required specifications. This includes electrical testing to check for shorts and opens, as well as inspection methods like Automated Optical Inspection (AOI) and X-ray inspection for multi-layered boards.

The creation of PCBs is a highly technical process that requires precision and careful planning. Understanding PCB fundamentals is essential for electronic prototyping, as it ensures the successful integration of electronic components into a functional unit. This knowledge is not just crucial for designers and engineers but also for anyone involved in product development, as it shapes the decision-making process from design to mass production.

In the rapidly evolving field of electronics, PCB design and fabrication remain at the forefront, constantly adapting to new technologies and demands. Innovations in PCB materials, fabrication techniques, and component assembly methods continue to push the boundaries of what can be achieved in electronic prototyping, enabling more sophisticated and compact designs.

Transitioning from the fundamental understanding of PCB design and fabrication, we move towards the critical phase of translating electronic designs into physical PCB prototypes. This stage marks the convergence of theoretical electronic concepts and practical, tangible implementations. The journey from a schematic diagram to a working PCB prototype is both intricate and enlightening.

From Design to PCB: Translating Electronic Designs into Physical Prototypes

Schematic to Board Layout: The first step in this transition involves converting the electronic schematic into a PCB layout using specialized software. This layout is essentially a detailed map that dictates where each component will be placed on the board and how they will be interconnected through traces, or conductive pathways.

Component Placement and Routing: In this stage, careful consideration is given to the placement of components and the routing of traces. Designers must balance electrical efficiency, signal integrity, and ease of manufacturing. Factors like minimizing noise interference, avoiding heat buildup, and ensuring accessibility for soldering and testing are crucial.

Prototype Fabrication: Once the layout is finalized, the PCB prototype is fabricated. This involves processes like applying photoresist, UV exposure for trace patterning, etching, drilling, and finally, mounting components onto the board, either through Surface Mount Technology (SMT) or Through-Hole Technology (THT).

Use of Rapid Prototyping Techniques: Advancements in rapid prototyping techniques, such as 3D printing of PCBs or quick-turn PCB fabrication services, have significantly accelerated this phase. These methods allow for faster iterations, enabling quicker testing and refinement of designs.

Testing and Iteration: Refining PCB Prototypes

Initial Power-Up and Basic Testing: The first test typically involves a simple power-up to ensure that there are no shorts or power issues. Basic functionality tests are conducted to ensure that all components are operational and that the board behaves as expected.

Advanced Functional Testing: More comprehensive testing is conducted to verify that the prototype meets all the specifications of the design. This includes testing for frequency response, signal integrity, thermal performance, and other critical parameters based on the intended use of the PCB.

Iterative Process for Optimization: Prototyping is inherently iterative. Based on the results of initial testing, changes and refinements are made to the design. This could involve rerouting traces, changing components, or even revising the overall layout to address any issues discovered during testing.

User Testing and Feedback: For consumer-focused products, prototypes may undergo user testing to gather feedback on usability, form factor, and functionality. This feedback can lead to further design modifications to enhance user experience.

Environmental and Stress Testing: In some cases, especially for industrial or automotive applications, PCB prototypes are subjected to environmental and stress testing. These tests assess the board's performance under various temperature ranges, vibrations, and other physical stresses.

Compliance and Certification Testing: Finally, PCB prototypes often need to undergo compliance testing to meet industry standards and certifications. This is particularly important for products in regulated sectors like medical devices or telecommunications.

The process of translating electronic designs into physical PCB prototypes is a critical bridge between conceptual design and real-world application. It involves a meticulous blend of technical expertise, practical considerations, and creative problem-solving. This phase is not just about creating a physical representation of an electronic schematic but ensuring that the prototype is functionally robust, manufacturably efficient, and tailored to meet specific user needs and industry standards.

The testing and iteration phase is equally crucial, acting as a refining forge where prototypes are tested, problems are identified and solved, and improvements are implemented. This iterative cycle is essential for honing the design to perfection, ensuring the final product performs reliably in its intended environment.

Machining Techniques in Prototype Development

Moving from the realm of electronic prototyping with PCBs, we transition into another crucial aspect of prototype development: machining. While PCBs are the lifeblood of electronic devices, machining brings to life the tangible, often structural components of a product. This process involves various techniques, each suited to different materials and design specifications, and plays a pivotal role in creating high-quality, functional prototypes.

CNC Machining: CNC (Computer Numerical Control) machining stands as a cornerstone in modern prototype development. It involves the use of computer-controlled machine tools to shape a material into the desired form. CNC machining is prized for its precision, consistency, and ability to produce complex geometries. It is suitable for a wide range of materials, including metals, plastics, and composites.

Milling and Turning: Milling involves cutting and drilling materials using a rotating cylindrical tool, while turning, often done on a lathe, involves rotating the material against a cutting tool. Both techniques are integral to creating parts with intricate details and are often used in conjunction with CNC technology for enhanced precision.

Other Techniques: Beyond CNC machining, milling, and turning, other techniques like EDM (Electrical Discharge Machining), waterjet cutting, and laser cutting are also used in prototype development. Each technique has its unique advantages and is chosen based on the material, complexity of the design, and the required precision.

Automotive Industry – Prototype Development of Engine Components

A compelling example of the application of machining in prototype development can be found in the automotive industry, specifically in the development of engine components.

The development of a new engine or improvements to an existing engine often requires the creation of prototypes for various components like pistons, crankshafts, or cylinder heads. These parts are crucial for the engine's performance and require precise manufacturing.

In creating these prototypes, CNC machining plays a critical role. For instance, the prototype of a piston may start as a solid block of aluminum, which is then shaped into the final form using a CNC milling machine. The precision of CNC machining allows for the creation of complex geometries, such as the intricate contours and channels found in a piston.

Once machined, these prototypes undergo extensive testing. For engine components, this might include stress testing, thermal analysis, and performance testing in simulated engine conditions. The insights gained from these tests are crucial for refining the prototype.

Often, machining is used in conjunction with other prototyping techniques like additive manufacturing. For example, a prototype cylinder head might be initially printed using AM to verify the design before being machined to achieve the required material properties and tolerances.

In the automotive industry, the precision and versatility of machining are invaluable in creating functional prototypes that accurately replicate the performance of the final product. The ability to quickly and accurately produce these complex components significantly accelerates the development cycle, allowing for rapid iteration and refinement.

Material Considerations in Machining for Prototypes

In the intricate process of machining for prototype development, the selection of appropriate materials stands as a critical decision. This choice influences the machining process and impacts the functionality, durability, and overall success of the prototype. Understanding the nuances of material properties and their implications in machining is essential for creating prototypes that accurately meet design specifications and project goals.

Matching Material Properties with Prototype Function: The foremost consideration in material selection is aligning the material properties with the intended function of the prototype.

For instance, if the prototype needs to withstand high temperatures or stresses, materials with high thermal resistance and tensile strength, such as certain grades of steel or titanium, might be required. Conversely, for prototypes where weight is a concern, like in aerospace applications, lighter materials such as aluminum or magnesium alloys could be more suitable.

Machinability of Materials: The ease with which a material can be machined, known as its machinability, is a key factor. Materials like brass and certain plastics are known for their good machinability, allowing for faster machining times and finer finishes. On the other hand, harder materials like stainless steel or nickel alloys can be more challenging to machine, potentially increasing time and costs.

Cost Considerations: The cost of materials is a significant factor, especially in the prototyping phase where budget constraints are often prominent. While it's important to choose materials that meet functional requirements, balancing these needs with cost considerations is crucial. In some cases, less expensive alternatives that still fulfill the necessary criteria can be used for prototyping, keeping the project within budget without compromising on essential qualities.

Surface Finish and Aesthetic Qualities: Depending on the prototype's end use, the aesthetic aspect of the material may be important. For consumer-facing products, the appearance of the prototype can be as crucial as its functionality. Materials with a naturally appealing finish, or those that can be easily finished to a high standard, are often preferred in these cases.

Environmental and Longevity Factors: The environmental impact of materials, along with their longevity and recyclability, are increasingly important considerations. For prototypes destined for eco-friendly projects, or those that need to demonstrate sustainability, materials that are recyclable or have a lower environmental footprint are preferable.

Compatibility with Other Prototype Elements: The material chosen for machining must be compatible with other elements of the prototype, especially in multi-material projects. This includes considering thermal expansion rates, chemical reactions between different materials, and the physical bonding of different components.

Response to Post-Machining Processes: The material's behavior in response to post-machining processes like heat treatments, anodizing, or painting is also a crucial factor. Some materials may alter their properties or dimensions when subjected to these processes, which can affect the prototype's accuracy and performance.

Availability and Lead Times: Finally, the availability of the material and associated lead times can influence the decision. Rarer or more specialized materials might have longer lead times, which can delay the prototyping process.

Selecting the right material for machining in prototype development involves a delicate balance of functional requirements, machinability, cost, aesthetics, environmental impact, compatibility with other components, response to post-processing, and availability. Each of these factors plays a vital role in determining the most suitable material for a given prototype, ensuring that the final product not only meets design specifications but also aligns with project timelines, budgets, and overarching goals.

Integration with Other Methods

The realm of prototype development is marked by the seamless integration of various methodologies, each contributing its unique strengths to the final product. In this context, the combination of machining with other prototyping techniques becomes crucial, especially for complex prototypes that demand a multifaceted approach. This integration not only enhances the capabilities of each individual method but also results in prototypes that are more sophisticated, functional, and closer to the final product in terms of quality and performance.

Combining Machining with Other Prototyping Methods

Machining and Additive Manufacturing Synergy: One of the most prevalent integrations is between machining and additive manufacturing (AM). While AM excels at creating complex geometries and internal structures, machining brings precision and superior surface finishes. For instance, a prototype may involve an intricate internal structure created through 3D printing, followed by CNC machining to achieve precise external dimensions and a high-quality surface finish. This combination is particularly beneficial in aerospace and automotive industries where parts often require both complex geometries and tight tolerances.

Machining and Casting Integration: Another effective combination is machining with casting methods, such as vacuum casting or injection molding. Machining can be used to create the master patterns or molds with high precision, which are then used for casting multiple copies of a part. This approach is advantageous when producing small batches of prototypes, offering both the detail of machining and the replicability of casting.

Incorporating Electronics with Machined Parts: In prototypes that involve electronic components, such as consumer electronics or IoT devices, integrating machined parts with PCBs and electronic elements is crucial. This involves precise machining of enclosures and structural components that house the electronic parts, ensuring accurate fit and alignment. Such integration is pivotal in creating fully functional prototypes that accurately represent the end product.

Machining and Sheet Metal Fabrication: For prototypes requiring metal parts with both solid and sheet components, integrating machining with sheet metal fabrication is essential. While machining can produce solid, 3D components, sheet metal fabrication is used for creating flat or bent metal parts. This combination is frequently used in the development of prototypes for appliances, industrial equipment, and automotive parts.

Surface Treatments and Finishing Techniques: Beyond shaping and forming, prototypes often require various surface treatments and finishing techniques to achieve the desired aesthetic and functional properties. This includes integrating machining with processes like anodizing, powder coating, painting, or plating. These finishing processes not only improve the appearance of the prototype but can also enhance its durability and resistance to environmental factors.

Hybrid Prototyping for Complex Assemblies: In projects where the prototype is a complex assembly involving various components, a hybrid approach combining machining with other techniques becomes vital. This approach allows for each part of the assembly to be created using the most suitable method, whether it be machining, 3D printing, or even hand-crafting for specific custom details. The final assembly of these diverse components results in a prototype that accurately reflects the multifaceted nature of the final product.

Integration with Soft Tooling: Machining is often used in conjunction with soft tooling methods for rapid prototyping. Soft tooling, which involves creating molds from softer, less durable materials, can be quickly and cost-effectively produced using machining. These molds are then used for short-run injection molding or casting, allowing for rapid iteration and testing of design concepts.

Collaborative Prototyping for Multidisciplinary Products: In the development of multidisciplinary products, such as medical devices or robotics, the integration of machining with other prototyping methods encourages collaboration across different fields. For example, a robotic limb prototype might combine machined structural elements, 3D-printed joints, and custom-designed electronic components, requiring expertise from mechanical engineering, materials science, and electronics.

The integration of machining with other prototyping methods represents a holistic approach to prototype development, especially for complex projects. This collaborative and multidimensional strategy enables the creation of prototypes that are representative of the final product in form and function, embodying the innovative spirit of modern product development. By leveraging the strengths of each method and harmoniously combining them, developers can create prototypes that push the boundaries of what is possible, setting the stage for the successful realization of sophisticated products.

After meticulously integrating various prototyping methods to construct Alpha prototypes, the next crucial phase in product development is testing and refining these prototypes. Alpha prototypes, being the initial physical manifestations of a concept, require thorough evaluation to ensure they meet the intended design criteria and functional requirements. This phase of testing and refinement is pivotal, as it sets the stage for making necessary adjustments before proceeding to the more refined Beta prototypes.

Testing and Refining Alpha Prototypes

Setting Test Criteria for Alpha Prototypes

Defining Functional Objectives: The first step in setting test criteria is defining the functional objectives of the prototype. What are the key functionalities it must demonstrate? For a tech gadget, this might include operational speed, user interface responsiveness, or battery life. For a mechanical device, it might be strength, durability, or efficiency of motion.

Establishing Performance Benchmarks: It's essential to establish clear performance benchmarks. These benchmarks act as a standard against which the prototype's performance can be measured. They should be realistic, attainable, and aligned with the end goals of the product.

Safety and Compliance Testing: Depending on the nature of the product, ensuring that the prototype meets industry safety standards and regulatory compliance is crucial. This is particularly important for products in regulated sectors like healthcare or automotive.

Identifying Key Metrics for Evaluation: Determine the key metrics for evaluating the prototype. This could range from quantitative data like speed measurements and load capacity to qualitative aspects like user experience and aesthetics.

The Process of Testing and Refinement

Initial Testing and Data Collection: Begin with a series of initial tests to collect data on how the prototype performs against the set criteria. These tests should cover a wide range of functions and scenarios to thoroughly evaluate the prototype's capabilities.

Analyzing Test Results: Once the data is collected, analyze it to identify areas where the prototype meets, exceeds, or falls short of expectations. This analysis is crucial for understanding the prototype's strengths and weaknesses.

Iterative Refinement: Use the insights gained from the analysis to make iterative refinements to the prototype. This might involve tweaking the design, adjusting materials, or modifying the manufacturing process. The goal is to enhance the prototype's performance and functionality based on real-world feedback.

User Feedback Integration: If the prototype is user-focused, integrating feedback from potential users or focus groups is invaluable. Their insights can provide practical perspectives on usability, design appeal, and functionality.

Repeated Testing Cycles: The testing and refinement process is typically iterative. Multiple cycles of testing, analysis, and refinement are common until the prototype meets all the predefined criteria and benchmarks.

Documentation of Changes and Results: Throughout the testing and refinement process, maintaining detailed documentation of changes made, test results, and feedback received is essential. This documentation provides a clear history of the prototype's evolution and informs future development phases and potential production.

By setting clear test criteria, rigorously evaluating performance, and systematically refining the design, developers can ensure that their prototypes are well-aligned with their vision and the project's objectives. This phase of development entails optimizing and enhancing the prototype, pushing it closer to the envisioned end product.

The insights gained during this phase are invaluable, as they often reveal unforeseen challenges or new opportunities for innovation. This continuous cycle of testing, analysis, and refinement ensures that the Alpha prototype evolves into a robust, functional, and user-friendly product.

The journey of an Alpha prototype from conception to refinement is often filled with challenges, insights, and breakthroughs.

Case Study: The Development and Testing of Dyson's Cordless Vacuum Cleaners

A notable example in the realm of Alpha prototype development is the work of Dyson, particularly in their development of cordless vacuum cleaners.

Initial Concept and Design Challenges: Dyson's venture into cordless vacuum cleaners was driven by the ambition to combine powerful suction with the convenience of a battery-operated device. The initial design challenges included creating a motor that was both compact and powerful, and a battery that could sustain long periods of usage.

Alpha Prototype Creation: The first step was the creation of Alpha prototypes to test these key components. The initial prototypes were far from the sleek designs seen in the final products. They were bulkier, used for internal testing to evaluate the core technology of the vacuum – the motor, and the battery life.

Iterative Testing and Refinement: Dyson's development process involved rigorous testing and iterative refinement. The Alpha prototypes underwent multiple rounds of performance testing, focusing on suction efficiency, battery life, and durability. Each round of testing provided feedback that was used to improve subsequent prototypes. This iterative process was crucial in gradually evolving the design into a more efficient and user-friendly form.

User-Centric Approach to Testing: Alongside technical performance testing, Dyson also focused on user experience with its Alpha prototypes. Early models were used in real-world environments to gather feedback on usability, ergonomics, and overall user satisfaction. This included evaluating the ease of handling, the efficiency of cleaning different surfaces, and the user interface.

Innovation in Motor Technology: One of the significant breakthroughs in the Alpha testing phase was the development of the Dyson digital motor. This motor was smaller and lighter than traditional motors but delivered the same level of power and efficiency. The Alpha prototypes were crucial in testing and perfecting this motor, ensuring that it could deliver consistent power without significantly draining the battery.

Battery Life Optimization: Another critical aspect of the development process was optimizing battery life. Early prototypes faced challenges with sustaining long periods of operation. Through iterative testing and refinement, Dyson engineers were able to enhance battery performance, ensuring that the vacuum cleaner could operate for a reasonable time on a single charge.

Addressing User Feedback: The feedback from users during the Alpha phase led to several design changes. For example, the placement of buttons and the balance of the vacuum cleaner were adjusted to make it more comfortable and intuitive to use. This user feedback was instrumental in guiding the design choices that would define the final product.

From Alpha to Beta Prototypes: The culmination of this extensive testing and refinement process was the transition from Alpha to Beta prototypes. These later prototypes were closer to the final product, with a more refined design that incorporated all the improvements and optimizations from the Alpha phase. They represented a significant step towards a market-ready product, showcasing the advancements in technology and design that had been achieved.

Final Product Launch: The final product, emerging from this rigorous development and testing process, was a line of cordless vacuum cleaners that set new standards in the industry. They were lighter, more powerful, and more efficient than previous models, and their success in the market was a testament to the thoroughness of the Alpha prototype development and testing process.

Dyson's journey in developing their cordless vacuum cleaners exemplifies the critical role of Alpha prototypes in product development. It highlights the importance of iterative testing, user feedback, and continuous refinement in turning a concept into a successful product.

Transitioning to Beta Prototypes

Following the rigorous process of Alpha prototype development and testing, as exemplified in Dyson's journey, we move into the subsequent critical phase of prototyping: the development and assessment of Beta prototypes. This stage represents a significant evolution from the Alpha stage, as the focus shifts from internal testing and refinement to external validation and user interaction.

Incorporating Alpha Feedback: The transition from Alpha to Beta prototypes involves integrating the feedback and data gathered during the Alpha testing phase. This includes addressing any identified issues, making design adjustments, and improving functionality based on real-world use and technical testing.

Enhanced Focus on Design and Usability: While Alpha prototypes might prioritize functionality, Beta prototypes place a greater emphasis on design, user experience, and usability. This stage is about refining the product to be more consumer-friendly, ensuring it not only works efficiently but also appeals aesthetically and is user-intuitive.

Preparing for Real-World Conditions: Beta prototypes are designed to withstand real-world conditions. This means they are often built with the same materials and processes as the final product, ensuring that they accurately represent how the finished product will look, feel, and function.

User Testing and Feedback

Engaging with Potential Customers or Users: A key component of Beta testing is involving potential customers or end-users.

This external testing provides invaluable insights into how the product is perceived, used, and experienced in a real-world context.

Gathering Diverse User Feedback: Beta testing typically involves a diverse range of users to ensure a comprehensive understanding of how the product performs across different demographics and usage scenarios. This can highlight unforeseen issues or new opportunities to enhance the product.

Iterative Approach Based on User Feedback: Similar to the Alpha phase, Beta testing is iterative. Feedback from users is analyzed and used to make further improvements. This continuous loop of testing, feedback, and refinement helps in fine-tuning the prototype to best meet user needs and expectations.

Usability and Experience Testing: Beta prototypes undergo usability testing to assess the ease of use, understanding of the interface, and overall user experience. This testing often includes qualitative measures like user satisfaction and ease of learning, as well as quantitative data like task completion time.

Market Readiness Assessment: Beta prototypes provide an opportunity to assess the product's market readiness. This includes evaluating not only the product itself but also packaging, instructions, and any supporting materials to ensure the entire product package is ready for market launch.

Risk Assessment and Final Adjustments: Finally, the Beta phase involves a thorough risk assessment to identify any remaining issues that could affect the product's performance, safety, or market acceptance. This comprehensive evaluation ensures that any potential risks are mitigated before the product enters the mass production stage.

Final Validation and Go/No-Go Decision: The culmination of the Beta testing phase is a critical decision point where the product is either approved for mass production ('Go') or sent back for further refinement ('No-Go'). This decision is based on the totality of feedback and data collected, balancing the product's technical performance, user experience, and market potential.

Chapter 08

Pilot Build – Bridging Prototypes and Full-Scale Production

Imagine standing at the threshold of a new era, where the abstract sketches and models of your product are about to transition into a tangible, market-ready reality. This pivotal moment is embodied in the concept of the pilot build, a phase in product development that bridges the gap between the theoretical world of prototypes and the tangible realm of full-scale production.

Definition and Purpose of a Pilot Build

A pilot build is essentially the dress rehearsal for your product's grand debut. It's a small-scale production run that serves multiple critical purposes in the product development journey. The primary objective is to test the waters of mass production, ensuring that the leap from a prototype to thousands of units is not just possible but also practical and efficient.

At this stage, the product is created using the final design, materials, and manufacturing processes. It's an opportunity to identify any last-minute issues that might not have been evident during the prototyping phases. This step is crucial for confirming that the product can be manufactured at scale without compromising quality, functionality, or aesthetics.

Transition from Prototypes to Pilot

The journey from Alpha and Beta prototypes to a pilot build is a significant shift in focus. While Alpha and Beta prototypes concentrate on validating the design and functionality of the product, the pilot build zeroes in on its manufacturability and scalability. This stage is about replicating the product with the exact materials and manufacturing processes that will be used in full-scale production.

A compelling example of this transition can be seen in the development of a cutting-edge smartwatch. The Alpha and Beta prototypes of the smartwatch were primarily focused on refining its design, ensuring user-friendly interfaces, and testing its functionalities like fitness tracking, notifications, and battery life.

As the product moved into the pilot build phase, the focus shifted. The smartwatch was now produced using the intended manufacturing methods and materials – the same ones that would be used for mass production. This stage was crucial for identifying challenges such as the precision assembly of tiny electronic components, the durability of the screen under real-world conditions, and the efficiency of the assembly line processes.

During the pilot build, a small batch of smartwatches was produced, which revealed a critical issue with battery life when the device was operated at full capacity. This insight was invaluable, as it led to adjustments in both the hardware and software, ensuring that the final product met the promised specifications.

Setting Objectives: Outlining the Goals for a Successful Pilot Build

A successful pilot build is a critical step in product development, encompassing much more than the production of a small batch of items. It's a comprehensive process that prepares a product for full-scale production. Setting clear, well-defined objectives for this phase is essential for achieving actionable results and valuable insights. These objectives not only guide the process but also ensure the pilot build serves as a solid foundation for the subsequent stages of product development.

Validating the Manufacturing Process:

Objective: Confirm the outlined manufacturing process is practical, efficient, and capable of producing the product according to specifications.

Focus Areas: Evaluate assembly lines, machinery capabilities, and workflow efficiency. The primary aim is to identify and resolve any process bottlenecks or inefficiencies. Additionally, assess the integration of new technologies or methods that could enhance production efficiency.

Additional Goals: Ensure environmental compliance and sustainability in the manufacturing process. This includes evaluating waste management and energy usage.

Ensuring Product Quality and Consistency:

Objective: Guarantee each unit produced during the pilot build adheres to the quality benchmarks established in the prototyping phase.

Focus Areas: Implement comprehensive quality control protocols and conduct detailed testing on each unit. Emphasis should be placed on long-term durability and performance under varied conditions.

Additional Goals: Develop a framework for ongoing quality assessment beyond the pilot phase, laying the groundwork for continuous improvement in full-scale production.

Testing Supply Chain and Logistics:

Objective: Examine the effectiveness of the supply chain and logistics strategy.

Focus Areas: Confirm the timely delivery of materials, effective inventory management, and robustness of distribution logistics. Assess the resilience of the supply chain to disruptions or external challenges.

Additional Goals: Explore opportunities for optimizing logistics costs and evaluate alternative or backup supply chain strategies to mitigate potential risks.

Finalizing Cost Estimates for Mass Production:

Objective: Determine an accurate cost structure for mass production based on insights from the pilot build.

Focus Areas: This includes a thorough analysis of material costs, labor, overheads, and additional expenses experienced during the pilot phase. Evaluate opportunities for cost reduction without compromising quality.

Additional Goals: Establish a scalable cost model that accommodates potential future expansion or adjustments in production volume.

Gathering User Feedback on Pre-Production Units:

Objective: Collect and assimilate user feedback on the pilot batch to inform final adjustments before mass production.

Focus Areas: Distribute pilot units to a carefully selected group of users and gather detailed feedback on usability, design, and functionality. Pay attention to the user experience and any unexpected use cases that emerge.

Additional Goals: Develop a methodology for ongoing user feedback collection post-launch, to continually refine the product based on real-world usage.

Identifying Potential for Scalability and Process Improvement

The goal here is to bridge the pilot phase and full-scale production, focusing on the product's ability to adapt to larger production volumes without compromising quality or efficiency.

Objective: Investigate the scalability of the product from the pilot phase to mass production.

Focus Areas:

Efficiency Analysis: Evaluate the current production process for its capability to handle increased volumes. This includes assessing the capacity of assembly lines, machinery, and labor resources.

Process Streamlining: Identify any steps in the production process that could be streamlined or automated to enhance efficiency at a larger scale.

Cost-Volume Assessment: Conduct an analysis to understand how costs will vary with increased production volumes. This should include considerations for bulk purchasing of materials and potential economies of scale.

Supply Chain Scalability: Ensure that the supply chain is robust and flexible enough to handle scaled-up operations. This includes evaluating suppliers for their ability to meet increased demands.

Compliance and Safety Assurance

Maintaining compliance with safety and regulatory standards is crucial, particularly for products in regulated industries. This step is about safeguarding the product against legal and safety challenges.

Objective: Guarantee adherence to all relevant safety and regulatory standards in the product and its production process.

Focus Areas:

Regulatory Compliance Review: Conduct thorough reviews to ensure the product meets industry-specific regulations. This may involve third-party testing and certification processes.

Safety Testing: Implement comprehensive safety testing procedures, considering both short-term and long-term usage of the product.

Documentation and Record Keeping: Maintain detailed records of compliance and safety testing. This documentation is crucial for regulatory reviews and future audits.

Preparing for Market Launch

This stage involves finalizing the strategies and preparations for introducing the product to the market.

Objective: Lay the groundwork for a successful market introduction.

Focus Areas:

Marketing Strategy Development: Develop comprehensive marketing strategies that include targeted campaigns, product positioning, and promotional activities.

Packaging and Branding Finalization: Confirm that packaging design and branding elements are in place, ensuring they align with the overall marketing strategy and appeal to the target audience.

Distribution Network Preparation: Set up efficient distribution channels. This includes logistics planning, establishing relationships with distributors or retailers, and ensuring a smooth supply chain from the factory to the end consumer.

Customer Support Systems: Implement customer support systems, including service centers, helpdesks, and online support, to ensure a positive customer experience post-launch.

Identification and Selection of Manufacturing Partner for Pilot Build

Selecting the appropriate manufacturing partner for a pilot build is a critical decision that shapes the course of the product's initial journey into a tangible entity. This selection process is centered around finding a manufacturer, often referred to as an Electronic System Design & Manufacturing (ESDM) company or Original Equipment Manufacturer (OEM), who can accurately translate the prototyped concept into a limited run of pre-production units. This step becomes pivotal once the design is solidified based on rigorous testing and validation.

Communication between the startup and the chosen ESDM or OEM is facilitated through detailed documentation, ensuring that every nuance of the product design is thoroughly understood and adhered to. These pilot units, manufactured at the ESDM facility, are often considered the first batch of volume production, setting a precedent for full-scale manufacturing.

For startups opting to establish their own assembly lines, it's imperative to adhere to specific quality standards. This approach demands a meticulous setup process, ensuring that the assembly line not only meets but exceeds industry norms for quality and efficiency. The groundwork laid in this phase is crucial, as it directly impacts the scalability and quality consistency of the product in its subsequent volume manufacturing stages.

Criteria for Choosing a Manufacturer for Pilot Build

Specialization in Pilot Builds: Seek a manufacturer with specialized experience in pilot builds. This expertise ensures they are familiar with the unique challenges and requirements of transitioning from a prototype to a small-scale production run.

Quality Assurance for Small Batches: The manufacturer must have a strong quality assurance process specifically tailored for small-scale production. The ability to maintain high-quality standards even in limited production runs is crucial.

Flexibility and Adaptability: Choose a partner who exhibits flexibility and adaptability in their processes. Pilot builds often require quick adjustments and the manufacturer should be able to accommodate iterative changes based on testing and feedback.

Technical Capabilities and Equipment: Assess the manufacturer's technical capabilities and equipment. Ensure they have the necessary technology and machinery to produce the pilot units with the precision and quality required.

Communication and Collaboration: Effective communication is key. The ideal manufacturing partner should be collaborative, open to suggestions, and willing to engage in a transparent dialogue throughout the pilot build process.

Turnaround Time and Responsiveness: Given that pilot builds often precede critical market launches, select a manufacturer who can commit to and meet specific turnaround times. Their ability to respond quickly and efficiently to any arising issues is vital.

Cost-Effectiveness for Pilot Scale: Evaluate the cost-effectiveness of their services. For a pilot build, the economic feasibility of producing a smaller number of units is a significant consideration. The manufacturer should offer a pricing structure that aligns with the limited scale and budget of a pilot build.

Proven Track Record with Pilot Builds: Look for a manufacturer with a proven track record in successfully executing pilot builds. References or case studies can provide insights into their ability to deliver as per the requirements.

Capacity for Future Scale-Up: While the immediate need is for pilot build production, considering the manufacturer's capacity to handle future full-scale production is advantageous. This foresight can streamline the transition from pilot to mass production.

Intellectual Property Protection: Ensure that the manufacturing partner has protocols to protect your intellectual property. The sensitivity of sharing designs and prototypes makes IP protection a critical criterion.

After the rigorous vetting process, establishing a strong, collaborative partnership with the chosen manufacturer is crucial for the success of the pilot build. This relationship is more than a simple client-vendor interaction; it's a strategic alliance where both parties work towards the shared goal of transforming a prototype into a successful pre-production model.

Surface Mount Technology (SMT) in Pilot Production

In pilot production, especially in electronics, Surface Mount Technology (SMT) plays a pivotal role. SMT is a method for producing electronic circuits in which the components are mounted directly onto the surface of printed circuit boards (PCBs). Understanding the nuances of SMT in the context of pilot production is crucial, as it can significantly influence the efficiency, quality, and scalability of electronic product manufacturing.

Basics of SMT in Pilot Production

SMT is distinguished by its ability to streamline the assembly process and enhance the density of components. This technology is particularly advantageous in pilot builds for several reasons:

Miniaturization: SMT allows for smaller components, which is essential in modern electronic devices where compactness is a key attribute.

Increased Circuit Speeds: Smaller components mean shorter electrical paths, leading to increased circuit speeds, a critical factor in many electronic devices.

High Component Density: SMT accommodates more components per unit area on the PCB, essential for complex pilot projects that require advanced functionalities within limited space.

SMT Assembly Techniques and Best Practices

The SMT assembly process for pilot production involves several key steps and best practices:

PCB Design and Layout: The design phase is crucial, as it sets the foundation for the SMT process. It includes planning the layout of components on the PCB, ensuring optimal placement for both performance and manufacturability.

Solder Paste Application: A stencil is used to apply solder paste to specific areas of the PCB where components will be placed. The quality of solder paste and precision in application are vital to prevent issues like solder bridging or insufficient soldering.

Component Placement: SMT machines, or pick-and-place machines, are used to accurately place components on the PCB. For pilot builds, this process needs to be carefully monitored to ensure accuracy, as the placement machines may need adjustments to handle different component sizes or layouts.

Reflow Soldering: After placement, the PCBs pass through a reflow oven where the solder paste is heated and melted, forming solder joints. The temperature profile of the reflow process is critical and must be tailored to the specific solder and components used.

Inspection and Quality Control: Post-assembly inspection is crucial. Techniques like Automated Optical Inspection (AOI) are used to detect any defects or issues in soldering or component placement.

Pilot Production of Apple Watch

A concrete example of Surface Mount Technology (SMT) playing a pivotal role in pilot production can be seen in the development of the Apple Watch. This innovative smartwatch, known for its sleek design and multitude of functionalities, faced significant manufacturing challenges during its pilot production phase.

Challenge: The primary challenge in the Apple Watch's development was the integration of a multitude of high-precision components into an exceptionally compact printed circuit board (PCB). The device needed to incorporate advanced sensors for health monitoring, a sophisticated processor for smooth operation, and various connectivity modules, all within the confines of a small, wrist-worn device.

SMT Implementation: In the pilot build phase, Apple employed SMT to meticulously mount and solder the miniature components onto the PCB. The process demanded extreme precision to accommodate the Apple Watch's compact design without compromising on functionality or risking damage to the sensitive components.

Outcome: The successful application of SMT in the pilot production of the Apple Watch was a critical factor in confirming the feasibility of the product's design. It demonstrated that the watch could be manufactured to meet Apple's high standards of quality, functionality, and aesthetic appeal. This successful pilot build was crucial in paving the way for the Apple Watch's transition to full-scale production and subsequent launch into the global market.

The Apple Watch's pilot production exemplifies the significance of SMT in the manufacturing of sophisticated electronic devices. The precise implementation of SMT, from the initial stages of PCB design to the final stages of soldering and quality inspection, was essential in ensuring the pilot build's success. This case highlights how SMT facilitates the miniaturization and functional complexity required in modern electronic devices, especially in pilot production phases where each detail sets the precedent for mass production.

Quality Assurance in Surface Mount Technology Processes

Quality assurance in Surface Mount Technology (SMT) is a cornerstone for ensuring the reliability and functionality of electronic devices, especially during the pilot production phase. This phase is critical, as it sets the standards and benchmarks for full-scale production. Implementing rigorous quality assurance practices in SMT processes ensures that each unit produced meets the highest standards of quality and reliability.

Development of a Comprehensive QA Plan:

Establishing Standards: Develop a comprehensive quality assurance plan that aligns with industry standards and specific product requirements. This includes defining clear quality criteria for each component and solder joint.

Documentation: Maintain detailed documentation of QA processes and standards. This ensures consistency in quality assessment and aids in tracing any issues back to their source.

SMT Process Monitoring:

Real-time Monitoring: Implement real-time monitoring of the SMT process. This involves overseeing the solder paste application, component placement, and reflow soldering to identify and rectify any issues immediately.

Automated Optical Inspection (AOI): Utilize AOI systems post-reflow to check for solder quality, component placement, and polarity. This helps in detecting and addressing defects early in the process.

Regular Testing and Inspection:

In-line Testing: Conduct regular in-line testing during the SMT process. This includes electrical testing for component functionality and connectivity.

Post-Assembly Inspection: After SMT assembly, conduct thorough inspections, including manual checks and additional machine-assisted inspections like X-ray analysis, to ensure the integrity of solder joints, especially for fine-pitch or BGA components.

Statistical Process Control (SPC):

Data Analysis: Implement Statistical Process Control to analyze data from the SMT process. This helps in identifying trends, predicting potential issues, and implementing preventive measures.

Process Improvement: Use SPC data to continuously improve SMT processes, aiming for higher efficiency and reduced defect rates.

Handling and Storage Controls:

Component Handling: Establish strict handling protocols for SMT components to prevent damage and contamination.

Storage Conditions: Ensure components and PCBs are stored under appropriate conditions to prevent moisture absorption or other environmental damages that could affect solderability.

Employee Training and Expertise:

Skilled Personnel: Ensure that the staff operating SMT lines are well-trained and knowledgeable about the SMT process and QA standards.

Ongoing Training: Provide regular training and updates to the staff on the latest SMT techniques and quality control measures.

Feedback Integration and Continuous Improvement:

Feedback Loops: Create feedback loops to gather insights from the QA process and integrate these learnings into continuous improvements in the SMT process.

Adaptation to Changes: Stay adaptable to new technologies and methods in SMT, incorporating these advancements into the QA process for better outcomes.

Ensuring reliability and quality in SMT processes during pilot production is about meticulous planning, constant monitoring, regular testing, and ongoing process improvement. These steps are fundamental in producing electronic components and devices that are functional, durable and reliable, setting a high standard for subsequent full-scale production.

Ball Grid Array (BGA) Assembly in Pilot Production

In pilot production, especially for sophisticated electronic devices, Ball Grid Array (BGA) assembly plays a crucial role. Understanding

BGA and its specific application in circuit board assembly during the pilot build phase is essential for ensuring high-quality, reliable electronic products.

Understanding BGA: An Introduction

BGA is a type of surface-mount packaging used for integrated circuits. It is distinguished by its use of tiny solder balls, which are arranged in a grid pattern on the underside of the package. When heated, these solder balls melt to form reliable electrical and mechanical connections to the PCB.

Applications in Circuit Board Assembly:

Compact and Efficient: BGA packages are noted for their compact size, making them ideal for devices where space is at a premium. This efficiency is particularly beneficial in pilot builds where the integration of multiple functionalities into a small form factor is often tested.

High Performance: BGAs offer superior performance, especially in terms of electrical conductivity and heat dissipation. This makes them suitable for high-performance circuits that are commonly prototyped in pilot builds.

Precision and Technique in BGA Assembly:

Accurate Placement: The process of placing BGA packages onto PCBs demands extreme precision. In pilot production, where each unit is crucial, the accuracy of BGA placement directly impacts the overall quality of the product.

Advanced Equipment: Utilizing advanced pick-and-place machines capable of handling the delicate nature of BGA packages is essential. This equipment must be calibrated specifically for the small size and precise placement requirements of BGAs.

Soldering and Reflow Process:

Controlled Reflow Soldering: The soldering of BGA packages requires a carefully controlled reflow process. The temperature profile needs to be meticulously managed to ensure even heating and prevent defects like bridging or cold solder joints.

Inspection Techniques: Due to the hidden nature of BGA solder joints, specialized inspection techniques such as X-ray inspection are employed in pilot builds. This allows for the examination of solder connections beneath the BGA package, ensuring their integrity and quality.

Rework and Repairability:

Challenges in Rework: Repairing or reworking BGA packages is challenging due to the inaccessible solder joints. In pilot production, developing techniques for efficient rework can save costs and time.

Specialized Rework Stations: Utilizing specialized rework stations that can accurately target heat to individual BGA packages is crucial. This allows for precise repairs without affecting adjacent components on the PCB.

Reliability Testing:

Stress Testing: BGAs in pilot builds undergo rigorous stress testing to evaluate their reliability under various conditions, including thermal cycling, vibration, and humidity exposure.

Failure Analysis: If failures occur, conducting a thorough failure analysis is key to understanding the root causes and refining the BGA assembly process for future builds.

BGA Assembly Process: A Step-by-Step Guide

The Ball Grid Array (BGA) assembly process is a critical component in the manufacturing of modern electronic devices, particularly during the pilot production phase. This process requires precision and expertise to ensure high-quality results. Here's a detailed step-by-step breakdown of the BGA assembly process:

PCB Preparation:

Cleaning: The printed circuit board (PCB) is thoroughly cleaned to remove any contaminants that could affect solder adhesion.

Inspection: The PCB is inspected for any defects that might impact the assembly process, such as cracks or warping.

Solder Paste Application:

Stencil Positioning: A stencil is precisely aligned over the PCB to apply solder paste only to the specific areas where the BGA will be placed.

Solder Paste Deposition: A controlled amount of solder paste is deposited through the stencil onto the PCB. The quality of solder paste and the accuracy of its application are crucial for strong and reliable solder joints.

BGA Placement:

Pick-and-Place Machine: The BGA package is picked up by a specialized pick-and-place machine equipped with high-precision cameras and alignment systems.

Accurate Placement: The BGA is accurately placed on the prepared area with solder paste. Precise alignment is essential to ensure that each solder ball aligns with the corresponding pad on the PCB.

Reflow Soldering:

Reflow Oven: The PCB with the placed BGA is passed through a reflow oven. The solder paste is heated to a specific temperature profile, causing it to melt and form solder joints.

Cooling Phase: Post-soldering, the PCB is gradually cooled to solidify the solder, forming robust electrical and mechanical connections.

Inspection and Quality Control:

X-Ray Inspection: Due to the BGA's hidden solder joints, X-ray inspection is employed to assess the quality of the solder connections beneath the BGA package.

Automated Optical Inspection (AOI): Further inspection may include AOI to check for any alignment issues or surface-level defects.

Rework if Necessary:

BGA Rework Stations: If defects are detected during inspection, specialized BGA rework stations are used. These stations can precisely heat and remove the BGA, allowing for the correction of issues.

Re-application Process: After rework, the BGA is re-applied following the initial steps, ensuring the integrity of the reworked connections.

Final Functional Testing:

Electrical Testing: The assembled PCB undergoes electrical testing to ensure that the BGA and other components function as intended.

Stress Testing: Additional stress tests may be conducted to assess the durability and reliability of the BGA connections under various environmental conditions.

Each step, from PCB preparation to final testing, is carried out with utmost precision to ensure the highest quality of the assembled product. This detailed approach in the pilot phase is crucial in identifying potential issues and optimizing the process for subsequent full-scale production, ensuring that the final products are reliable and meet the stringent standards of modern electronic devices.

Challenges and Solutions in BGA Assembly for Pilot Builds

Ball Grid Array (BGA) assembly, while efficient and space-saving, presents unique challenges, particularly in the context of pilot builds. Addressing these issues effectively is crucial for ensuring the success and reliability of the final product. Let's explore some common challenges encountered in BGA assembly during pilot builds and the corresponding solutions.

Misalignment of BGA Components:

Challenge: One of the most common issues in BGA assembly is the misalignment of components, where the solder balls do not perfectly align with the PCB pads.

Solution: Enhancing the precision of pick-and-place machines and using more sophisticated alignment technologies can mitigate this issue. Regular calibration and maintenance of equipment are also vital.

Inadequate Solder Joint Formation:

Challenge: Inadequate or incomplete solder joint formation can lead to weak connections, affecting the functionality of the device.

Solution: Optimizing the reflow soldering process by fine-tuning the temperature profile is key. Ensuring the quality of solder paste and its precise application also plays a crucial role.

Thermal Stress and Warping:

Challenge: BGA components are sensitive to thermal stress, which can cause warping or damage during the soldering process.

Solution: Implementing a controlled and gradual heating and cooling process in the reflow oven helps minimize thermal stress. Using materials and PCB designs that are less prone to warping can also be beneficial.

Solder Bridging:

Challenge: Solder bridging occurs when solder connects two adjacent pads, creating an unintended connection.

Solution: Precise control of solder paste volume and refinement of stencil design can reduce the occurrence of solder bridging. Enhanced inspection methods like AOI and X-ray can help detect and address this issue early in the process.

Defective Solder Balls:

Challenge: Defective or damaged solder balls can compromise the integrity of the solder joints.

Solution: Implementing stringent quality control measures for solder balls and increasing the inspection frequency can help identify and rectify this issue.

BGA Rework Challenges:

Challenge: Reworking BGA components, especially in complex PCBs, can be challenging and risks damaging the board.

Solution: Using specialized BGA rework stations and skilled technicians for rework tasks can ensure effective and safe repairs. Developing a rework protocol that minimizes risks to other components is also important.

Imagine a scenario where a tech company is in the pilot build phase of a new smartwatch. They encounter a recurring issue with the misalignment of BGA components on the PCB, leading to inconsistent device performance.

To tackle this, the company revises its assembly process. They recalibrate their pick-and-place machines for greater precision and implement an

advanced optical alignment system that offers higher accuracy. Additionally, they introduce more frequent checks and calibrations of their assembly equipment to maintain this precision consistently.

In parallel, they optimize their reflow soldering profile, conducting a series of tests to identify the ideal temperature curve that reduces the risk of misalignment due to thermal expansion. They also enhance their post-soldering inspection process, incorporating a combination of AOI and X-ray inspection to identify and rectify any misalignments immediately after soldering.

The result of these changes is a marked improvement in the quality and consistency of the BGA assembly for their smartwatch pilot builds. The instances of misalignment drop significantly, leading to a higher yield of fully functional devices. This meticulous approach to addressing the misalignment issue not only improves the current pilot build but also sets a higher standard for future production runs, ensuring long-term reliability and customer satisfaction.

Injection Molding and Machining for Component Fabrication in Pilot Builds

In the pilot build phase of product development, component fabrication often involves specialized processes like injection molding and machining. These techniques are crucial for creating high-quality, precise parts that are essential for the assembly and functionality of the final product.

Injection Molding Process in Pilot Builds

Injection molding is a popular method for manufacturing plastic parts, and its role in pilot builds is significant due to its efficiency and ability to produce complex shapes with high precision.

Mold Design and Creation: The process begins with designing and creating a mold. For pilot builds, this often involves a careful balance between the complexity of the design and the cost. Rapid tooling techniques might be employed to create molds faster and more cost-effectively than traditional methods.

Material Selection: Choosing the right plastic material is crucial. For pilot builds, materials are selected based on their properties, such as strength, flexibility, and resistance to temperature or chemicals, as well as how they interact with the mold.

The Injection Molding Process: The selected plastic material is melted and then injected into the mold under high pressure. This process is precisely controlled to ensure consistent quality and to avoid defects like air pockets or warping.

Cooling and Ejection: After injection, the material cools and solidifies within the mold. Once solidified, the part is ejected. The rapid cooling time in injection molding makes it an efficient process for pilot builds.

Quality Control: Each part is inspected for quality. In pilot builds, this inspection is critical to ensure that every component meets the necessary specifications before moving to assembly.

Machining in Pilot Builds

Machining offers unparalleled precision for creating metal parts and is a vital component in pilot builds, especially for parts that require exact tolerances and specific finishes.

CNC Machining: Computer Numerical Control (CNC) machining is commonly used in pilot builds. It involves using computer-controlled machine tools to shape the metal. CNC machining is preferred for its accuracy and repeatability, which are essential in pilot builds.

Material Selection: Similar to injection molding, selecting the right material for machining is crucial. Metals are chosen based on the part's required strength, durability, and other physical properties.

Process Precision: Machining in pilot builds often requires creating complex geometries with tight tolerances. Precision is paramount to ensure that each part fits perfectly within the product.

Prototyping to Production Transition: Machining in pilot builds allows for easy transition from prototyping to production. Adjustments to designs can be made rapidly, ensuring that the transition to full-scale production is as smooth as possible.

Quality and Inspection: Each machined part undergoes rigorous inspection. This includes dimensional accuracy, surface finish, and ensuring the part meets all design specifications.

Material Considerations for Injection Molding and Machining in Pilot Builds

Selecting the right materials for injection molding and machining during the pilot build phase is a critical decision that significantly impacts the success of the product. This choice influences the efficiency and effectiveness of the manufacturing process

Material Selection for Injection Moulding

Thermoplastics Selection: The most commonly used materials in injection molding are thermoplastics. For pilot builds, the selection is typically based on the material's properties such as strength, flexibility, heat resistance, and cost-effectiveness. Common choices include ABS (Acrylonitrile Butadiene Styrene), polycarbonate, and polyamides (nylons).

Prototyping vs. End-Use Material: In pilot builds, the material chosen for injection molding might differ from the one intended for full-scale production, especially if cost or availability is a concern.

It's important to choose a material that behaves similarly to the end-use material to ensure accurate testing and validation.

Mold Compatibility: The chosen material must be compatible with the mold design. Factors like shrinkage rate, flow rate, and cooling time must be considered to avoid defects such as warping, sink marks, or incomplete filling.

Surface Finish and Color: Material selection also impacts the surface finish and color of the final product. Some materials take paint and textures better than others, which is crucial for products where aesthetics are important.

Material Selection for Machining

Metal Selection: For machining, metals are chosen based on their machinability, strength, and cost. Aluminum is popular for its ease of machining and lightweight, while stainless steel is chosen for its strength and corrosion resistance. For more specialized applications, materials like titanium may be used.

Tolerances and Precision: The material's ability to maintain tight tolerances and precision is key. Softer metals might be easier to machine but can deform easily, affecting precision. Conversely, harder metals offer precision but can be more challenging and time-consuming to machine.

Thermal Properties: Consider the thermal properties of the material, especially if the component will be subjected to high temperatures. Materials with high thermal stability are essential in such cases to prevent deformation or loss of mechanical properties.

Post-Processing Requirements: Materials that require minimal post-processing (like polishing or finishing) can reduce overall production time and cost in pilot builds. The ease of post-processing also affects the speed of iteration and refinement in the pilot phase.

Prototype to Production Transition: Choose materials that allow for a smooth transition from pilot build to full-scale production. Consider the availability and cost of the material at a larger scale, as well as any changes in machining requirements.

Material selection for injection molding and machining in the context of a pilot build requires a careful balance of functional, aesthetic, and production considerations. The choice of materials influences the quality and performance of the prototype, ensuring the feasibility and scalability of the manufacturing process.

Testing Procedures in Pilot Build

Testing procedures during the pilot build phase are crucial for ensuring that the product meets all design specifications and functions as intended before it moves into full-scale production. A robust testing framework in this stage not only validates the product's design and performance but also identifies areas for improvement.

Performance and Durability Tests in Pilot Build

In the pilot build phase, performance and durability tests are vital to ensure that the product not only functions according to its design but also stands the test of time and usage. These tests are designed to rigorously challenge the product under various conditions, providing insights into its longevity and reliability.

Performance Testing:

Functionality Checks: This involves testing the product to ensure it performs all its intended functions correctly. For electronic devices, this might include power-up tests, software functionality checks, and connectivity tests.

Load Testing: Assess the product's performance under maximum operational capacity. This helps identify potential points of failure under stress.

Speed and Responsiveness: For products where speed is a factor (like computing devices), tests are conducted to measure response times and processing speeds under different scenarios.

Durability Testing:

Lifecycle Testing: Simulate the entire lifecycle of the product to assess how it holds up over time. This can involve accelerated life testing techniques where the product is subjected to the equivalent of months or years of use in a shortened period.

Environmental Exposure: Expose the product to various environmental conditions, including temperature extremes, humidity, dust, and UV exposure, to ensure it can withstand diverse environments.

Mechanical Stress Tests: Perform tests like drop tests, vibration tests, and shock tests to evaluate the product's physical robustness.

Specialized Performance Tests:

Energy Efficiency Tests: For products that use power, testing for energy efficiency and battery life is crucial. This includes monitoring power consumption patterns and battery endurance under different usage scenarios.

Compatibility Testing: Ensure that the product is compatible with other systems or devices it's intended to work with. This is particularly important for tech products that are part of a larger ecosystem.

User-Centric Performance Assessments:

User Interface Testing: For products with a user interface, conduct tests to ensure that the interface is intuitive, responsive, and user-friendly.

Real-World Usage Scenarios: Test the product in scenarios that mimic how it will be used in real life to assess its performance in practical settings.

Continuous Monitoring and Data Collection:

Data-Driven Insights: Throughout the testing process, collect and analyze data to gain insights into performance trends and potential areas of improvement.

Iterative Improvements: Use the data collected to make iterative improvements to the product, enhancing its performance and durability.

By conducting comprehensive performance and durability tests during the pilot build phase, manufacturers can identify and rectify potential issues before the product moves to full-scale production.

Compliance and Safety Testing in Pilot Builds

Compliance and safety testing during the pilot build ensures that the product adheres to various industry standards and safety regulations. This testing is a fundamental aspect of product development that safeguards the end-user and enhances the product's market viability.

Identification of Relevant Standards and Regulations:

Research and Analysis: Begin by identifying all relevant standards and regulations applicable to the product. This includes industry-specific safety standards, environmental regulations, and consumer protection laws.

International and Regional Compliance: For products intended for international markets, ensure compliance with regulations across different regions, which may have varying standards.

Compliance Testing Protocols:

Developing Testing Procedures: Establish testing procedures that precisely align with the required standards. This might involve simulations, laboratory testing, and certification processes.

Engaging with Certified Labs: Often, compliance testing is conducted in collaboration with certified laboratories that specialize in regulatory testing. Their expertise ensures that the testing is thorough and meets all regulatory requirements.

Safety Testing:

Physical and Electrical Safety: Conduct tests for physical and electrical safety. For electronic products, this includes testing for electrical hazards, fire safety, and electromagnetic compatibility.

Material Safety: Test materials used in the product for toxicity, allergens, or any other hazards that could pose risks to users.

Specialized Industry-Specific Testing:

Medical Devices: For medical devices, compliance testing includes biocompatibility testing and validation of sterilization processes.

Children's Products: For products intended for children, additional safety tests are conducted, focusing on choking hazards, sharp edges, and toxic materials.

Documentation and Record Keeping:

Maintaining Compliance Records: Keep comprehensive records of all compliance testing results. This documentation is crucial for audits, certifications, and any legal requirements.

Updating Compliance Files: Regularly update compliance files to reflect any changes in regulations or standards.

Scaling from Pilot to Full-Scale Production

Transitioning from a pilot build to full-scale production is a critical phase in the product development lifecycle. It involves scaling up the manufacturing process while maintaining the quality, efficiency, and integrity achieved during the pilot phase. This stage requires strategic planning and careful management of various aspects, particularly in terms of production scalability and supply chain readiness.

Planning for Scale: Strategies for Transition

Analyzing Pilot Build Data: Start by thoroughly analyzing data and feedback from the pilot build. This includes assessing the success of the design, manufacturing process, and the product's performance. Identify any areas that need refinement before scaling up.

Modifying Design for Manufacturability: Based on the pilot build insights, make necessary design modifications to optimize manufacturability. This might include simplifying the design, choosing alternative materials, or tweaking the dimensions for easier and more cost-effective mass production.

Equipment and Facility Expansion: Evaluate the need for additional equipment or facility expansion to meet the demands of full-scale production. This includes assessing the capacity of existing machinery and the potential need for automation to improve efficiency.

Workforce Training and Scaling: Prepare for an expanded workforce if necessary. This involves not only hiring more personnel but also ensuring adequate training is provided to maintain the quality standards established in the pilot build.

Managing Supply Chain for Full-Scale Manufacturing

Supplier Evaluation and Engagement: Re-evaluate existing suppliers and engage new ones if needed to meet the increased demand. Ensure that suppliers are capable of delivering larger quantities without compromising on quality or lead time.

Building Inventory and Buffer Stocks: Develop a strategy for inventory management. This includes building buffer stocks to mitigate the risks of supply chain disruptions, which are more pronounced during full-scale production.

Logistics and Distribution Planning: Upgrade logistics and distribution plans to handle increased production volumes. This involves optimizing transportation routes, warehouse management, and possibly collaborating with logistics partners for efficient distribution.

Quality Control at Scale: Implement quality control measures that can be scaled effectively. This may involve more automated inspection processes or enhanced quality checks at various stages of the supply chain.

Risk Management and Contingency Planning: Develop robust risk management strategies. This includes having contingency plans for potential supply chain disruptions, machinery breakdowns, or sudden changes in market demand.

Sustainability Considerations: As production scales, consider the environmental impact. Implement sustainable practices in manufacturing and supply chain management to maintain corporate responsibility and compliance with environmental regulations.

The transition from pilot to full-scale production requires a balance between the learnings from the pilot build and the demands of mass production. It's about scaling up efficiently while retaining the core values and quality that were established during the pilot phase.

Conclusion

In the words of Carl Sagan, "Somewhere, something incredible is waiting to be known." In the context of product development, that 'something incredible' is the next groundbreaking idea, the next transformative product, the next leap in how we improve lives through innovation. This journey, much like the exploration of the cosmos, is filled with unknowns, challenges, and boundless potential for discovery and innovation.

The pursuit of transforming a concept into a tangible product is a venture of both courage and creativity. It's a path that demands technical acumen and a deep understanding of the human experience. Every stage of this process, from the initial spark of ideation to the meticulous execution of a pilot build, is a testament to the power of human ingenuity and perseverance.

Innovation, the heartbeat of successful product development, entails continually adapting to change, learning from each setback, and persistently pushing the boundaries of what is possible. It's about finding solutions that resonate with the needs of society, solutions that are as practical as they are visionary.

Moreover, the journey underscores the significance of viewing product development as an interconnected ecosystem. This ecosystem thrives on collaboration, where diverse ideas and perspectives converge to create something truly exceptional. It's a symphony of skills, knowledge, and creativity, where each component plays a crucial role in the harmony of the final product.

Yet, the essence of this journey is not just in the creation of products but in the transformation they bring. It's about harnessing the power of technology and innovation to improve lives, to solve real-world problems, and to open doors to new possibilities. It's a continuous quest to exceed the expectations of an ever-changing world.

As we traverse through this dynamic territory, let's embrace the unknown with a sense of optimism and purpose. Each obstacle encountered is an opportunity to refine our approach, to innovate, and to grow.

In the spirit of exploration and discovery, let us step forward into the future with a resolve to keep innovating, learning, and pushing the boundaries. The incredible, as Sagan hinted, is not just out there waiting to be discovered – it's within each visionary, each entrepreneur, each innovator who dares to dream and turn those dreams into reality. The next chapter of human progress in technology and product development awaits, and it's ours to write.

Author Bio

Dr. K.R. Suresh Nair, an esteemed alumnus of IIT Bombay, a serial entrepreneur stands at the forefront of innovative product design and technology development. As the visionary founder of Design Alpha, created in collaboration with Social Alpha, he has dedicated himself to nurturing entrepreneurs in critical sectors such as health, environment, agriculture, and education. Design Alpha with a strategic investment, got transitioned to Amara Raja Design Alpha providing concept to product development services. Dr. Nair is also Founder of Biophoton Technologies engaged in healthcare products.

With an unwavering belief in the potent synergy between academia and industry, Dr. Nair emerges as a guiding force in the realm of new technology advancement. From his humble beginnings in the R&D Lab of the Ministry of IT to his tenure as Chief Technology Officer at NeST, Dr. Nair consistently spearheads innovative progress. His passion for knowledge dissemination knows no bounds, exemplified by his extensive contributions: 107 publications, 18 patents, two co-authored books, and over 200 technical reports. Dr. Suresh Nair stands not only as an innovator and mentor, but as a true luminary in the realm of technology and product development.